CELL WARS

Marshall Goldberg, M.D.

CELL WARS

The Immune System's Newest Weapons Against Cancer

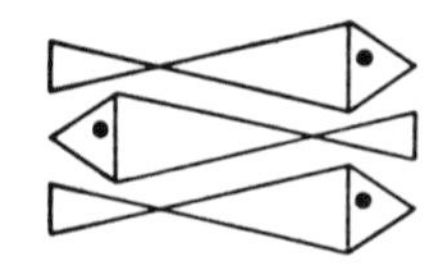

FARRAR · STRAUS · GIROUX

NEW YORK

Published simultaneously in Canada by
Collins Publishers, Toronto
Printed in the United States of America
Designed by Constance Fogler
First edition, 1988

Library of Congress Cataloging-in-Publication Data
Goldberg, Marshall.
Cell wars.
1. Antibodies, Monoclonal—Therapeutic use.
2. Cancer—Immunotherapy. I. Title.
RC271.M65G65 1988 616.99'406 88-7091

For

the scientists of

The Wistar Institute and

my wife, Barbara

Fortune is always on the side of the biggest battalions.

—*Madame de Sévigné* (1626–96)

CELL WARS

1

UNLESS I DIE SUDDENLY, I expect to face a big dilemma as the end of my life draws near. If I am reasonably certain no cure exists for my disease, I'd prefer to die at home. But if there's the slightest chance that a new treatment can save me at the last minute, I want to be where I can get it—which usually means an uncomfortable intensive-care unit. The same holds true for most advanced cancer victims. In different words and in different ways, they express the hope that some research lab will come up with a new drug, they will be the first to get it, and it will be so successful they will surprise everybody by walking out of the hospital.

It is not always a forlorn hope.

Similar cures were prayed for and received back in the late 1940s when the first big batches of penicillin arrived at American hospitals and patients with pneumonia, who probably would have died earlier, were saved.

Unfortunately, internal cancer, once it reaches a certain stage in development, is an exceedingly formidable foe. It hides in the midst of normal cells, it mutates into forms whose outward appearance is only slightly and subtly different from "self," it may even release substances into the bloodstream that sabotage those brain centers that regulate, or at least modulate, the functioning of the immune system. It can foil the body's defenses in any number of ways.

For these and other reasons, including stringent federal regulations, an effective new cancer treatment simply doesn't leap from the mind and test tubes of its creator to the clinic. Instead, it must go through three separate and distinct stages.

The first occurs in vitro: in a tissue-culture dish. Rapidly proliferating cancer cells are exposed to the new agent and its effects on their pattern of growth are observed. Sometimes a DNA building block, such as thymidine, is chemically linked to the radioactive isotope tritium and its uptake by the cancer cells used to gauge the extent of growth inhibition. At the same time, the new agent is tested for toxic effects on such essential body tissues as those of the liver and kidneys.

The next stage of testing is in vivo: in the living animal. The mouse, because many of its biological systems more closely resemble man's than do those of higher animals, is usually used for this purpose. A favorite is a scrawny little rodent called the "nude mouse" because it is hairless. It is also missing a thymus gland, the source of a key component of the immune system, the T-helper cell,* and so cannot easily reject a human tumor implanted under its skin. If the new agent destroys some or all of the mouse's tumor implant without doing damage to healthy tissues, its testing proceeds to the third stage: human experimentation.

Very few, however, make it this far.

This book, in part, is the story of one new treatment

* See Glossary.

method—monoclonal antibodies—that did. Conceptually at least, it may well represent the forerunner of a cure for most human cancers.

Before its death by suicide, PD-733 had been one of the thousands of cells lining the pancreatic duct of its human host. In appearance, it rather resembled a fortified port city—walled on three sides by similar cells, with a body of water, the duct, on the fourth. Deep in its interior, at the center of the cell, was its nuclear headquarters containing its control systems and master units. Between the outer boundaries and the nucleus stood numerous factories to convert raw nutrients into high-energy fuels and assembly plants to link amino acids into useful proteins.

Though for most of its length the cell's outer walls were separate from those of its neighbors, they came together at certain junctures called anchorage points and were further interconnected by a series of tunnels, bridges, and checkpoints. Its port side contained multiple docking facilities for the loading and unloading of intercellular vessels and a network of capsular relays to transport materials and message units rapidly from the docks to its mitochondrial factories and nuclear headquarters. In addition to the tunnels linking it to its neighbors, its interior was crisscrossed by roads and cablelike structures that served to move the cell short distances on caterpillarlike treads.

PD-733 could not think in the usual sense of the word. Neither could it innovate. But it did have bio-intelligence. Through a process of electromagnetic fluxes, it possessed a sense of order, of harmony with the whole, and could respond to certain nervous or hormonal stimuli in binary, yes-or-no, fashion. It was endowed with two basic functions: first, to manufacture and release into the pancreatic duct a digestive enzyme to break down the complex starches and sugars its host ingested and to ship them to other cells for use as building blocks or fuel; second, to replicate itself at predetermined intervals. Since coming into existence, PD-733 had divided twenty-eight times and was programmed for twenty more divisions before its genetic machinery ceased and it slipped into senescence. Such replications were a complicated process, necessitating much planning, coordination, and gathering together of materials. Only after these preparations were complete did the master control unit give the signal to begin cell division. Usually an internal mechanism determined its timing, but every once in a while, in emergency repair situations, the go-ahead came from the outside in the form of an urgent chemical message that, once received, superseded all previous orders and schedules. Even then, it was left to the master control unit to issue the ultimate signal, for a mistake in timing or coordination of efforts could so disrupt the process that, instead of identical daughter cells, monstrous mutants were produced. This had never happened to PD-733, but at the moment its master unit was in a state of confusion. The cell on its western border was behaving oddly, disharmoniously. It had abandoned its anchorage points and closed the tunnels and bridges used to transport essential nutrients and metals from the capillary channel on its far side to its neighbors. Instead of sharing such materials, it was confiscating them, expanding in size, and pushing outward as if about to divide—even though it had done this just recently and was not supposed to repeat the process for many months. Even more disturbing, the western cell was sending signal after signal for PD-733 to move out of its way. The master control unit refused to order this migration, sensing it would be disruptive and potentially disastrous to the tissue as a whole. Yet despite the flurry of messages sent back to the western cell, it continued to expand, putting the master unit in a quandary. Something had to be done. Its neighbor's aggressive behavior was not only depriving PD-733's factories of raw materials but impairing other critical functions all along their common border. Unable to improvise, the master unit had but two choices: to prepare for emergency cell division itself, regardless of the consequences to its neighbors, above and below, or to dissolve and destroy itself by releasing the digestive enzymes stored in huge internal vats. PD-733 didn't know—was incapable of knowing—what had gone wrong in the control centers of the cell next to it. Nor could it have acted to destroy that cell before being destroyed by it. Special immune system surveillance cells, macrophages or Natural Killer cells, should already have done that. But somehow the western cell had disguised or transformed itself sufficiently to elude their clutches, and PD-733 lacked the wherewithal to alert them. In lieu of this, it took what it perceived to be the proper step to preserve tissue harmony: it committed suicide.

And after many such internal episodes, its host, a man named James Federline, came to realize he was dying.

On August 30, 1985, my team at Flint's Hurley Medical Center treated its first cancer patient with The Wistar Institute's anti-gastrointestinal cancer monoclonal antibody.

Founded in 1892, Philadelphia's Wistar Institute of Anatomy and Biology is the oldest independent biomedical research facility in the United States. Its director, Hilary Koprowski, is the virologist who created and field-tested the first oral polio vaccine and is also one of the preeminent authorities on cancer-related immunological research in the world. The specific antibody, created by Drs. Koprowski and Zenon Steplewski at The

Wistar Institute in 1979, is called 17-1-A. As the first pure antitumor antibody in existence, it has found wide application in the detection of gastrointestinal cancers by various isotope tests, and its potential role in their treatment is currently being evaluated by eight medical centers worldwide.

In mid-November of 1985, by which time we had treated eight patients with this antibody, I got a phone call from The Wistar Institute's Zenon Steplewski asking me to accept into our program a man from the Washington, D.C., area who was suffering from pancreatic cancer. His case was complicated, Zenon warned, and as he summarized it I grew less and less enthusiastic. The patient had just returned home from the M. D. Anderson Hospital in Houston after receiving a six-week course of treatment with Tumor Necrosis Factor, a new and highly experimental anti-cancer agent. Yet his latest X-rays showed that the size of his tumor—huge to begin with—had not changed. Now, more or less as a last resort, he was requesting monoclonal antibody therapy. The man was very intelligent and well informed, Zenon said, and was the owner of a large construction company. Moreover, a member of The Wistar's board of managers had interceded on his behalf. As a favor to the Institute, would I consent to treat him?

In the end, after a brief conversation with the patient himself, I reluctantly agreed—for two reasons. By this time my hospital team owed The Wistar Institute not one favor but many for the thousands of dollars' worth of monoclonal antibodies and laboratory tests they had supplied us free of charge. And then there was the fact that Jim Federline was only forty-one years old.

A week later, Jim and his wife, Tinka, flew to Flint and appeared in my office.

Meeting a cancer victim, especially one in an advanced

stage of the disease, can be an unsettling experience. First comes the appraising glance (how bad off is he?). Then the handshake. And finally the moment when we look each other squarely in the eye.

Jim Federline was a dapper young man, thin but not gaunt, with a strong grip and a steady gaze. Sometimes I can detect skepticism or silent pleading in a cancer patient's scrutiny, but this time there was neither. I saw only relief: he was at last at a hospital that could give him the form of treatment he and his wife had sought from the start.

In my office, I explained the hospital procedure we would follow. Since the crucial preliminary step, to make sure The Wistar Institute antibody would selectively seek out and attack the cancer cells in a surgical specimen of his tumor, had already been done by Zenon Steplewski, we knew the treatment would be safe, simple, and speedy. No patient yet has suffered any significant side effects from his first monoclonal antibody infusion, and since we took elaborate precautions, we didn't expect any now. Apart from the minor discomfort of having a plastic catheter inserted into a large vein to connect his circulation to a blood-cell separator called a leukopheresis machine, the treatment was also painless. The leukopheresis procedure, taking around four hours, separated his red from his white blood cells, reinfused the red cells immediately, and isolated five to ten billion white blood cells (leukocytes) in a plastic bag. We would then incubate these potential killer cells with monoclonal antibodies that would instruct them to attack those cancer cells the 17-1-A antibodies had been bred to recognize, and infuse the mixture intravenously over one to two hours. If all went smoothly, Jim should be out of the hospital and on his way back home by late afternoon of the following day.

Left unmentioned was how successful the treatment might be. The Federlines didn't ask, and at this early stage in its testing, I wouldn't have known how to answer. The odds were heavily in favor of this man dying, and soon; the mean length of survival after the diagnosis of pancreatic cancer is a mere four months. The reason is that the pancreas gland is buried so deep in the center of the body that most cancers arising from it are silent killers. By the time their presence is even suspected, 90 percent of them are no longer surgically curable. In short, for all but a fortunate few, the diagnosis of pancreatic cancer is tantamount to a death sentence.

The Tumor Necrosis Factor infusions given Jim in Houston had failed to shrink his tumor. Could mouse-derived monoclonal antibodies—each less than a millionth the size of a cancer cell—do any better? Though the chances seemed slim, there were precedents: three of the first twenty gastrointestinal cancer patients treated at the Fox Chase Cancer Center in Philadelphia by this method responded favorably and one, harboring a similar, if smaller, pancreatic cancer, spectacularly.

Having heard of this one apparent cure, Jim and his wife hoped for the same result. But I had my doubts. In the coldly technical term cancer specialists use to explain their failures, his "tumor burden" was simply too great.

I did not get to know the Federlines very well during Jim's first treatment. Much as I admired their fortitude, their relentless search for a cure, I held back, expecting to treat Jim just once and, barring some miraculous response, never again. My personal involvement with them did not come until later and was fully shared by my research partner, Harland Verrill, as well as Zenon Steplewski and Hilary Koprowski of The Wistar Institute and many others. In time, Jim Federline came to be much more than another cancer patient—number nine in our Hurley Medical Center series; he became a cause and a source of wonder to us all. Inside his abdomen, a struggle raged between cancer cells and his immune defenses. On the basis of sheer numbers, the cancer should be winning, but it wasn't; Jim appeared to be getting better. We speculated that the monoclonal antibody treatment had given his immune system a boost, though none of us believed that was the whole story. Something more—some inner resource of healing power—was keeping his cancer in check, and as scientists, and fellow mortals, we were eager to learn more about its nature, its origin.

2

WHAT EXACTLY is an antibody? It is a Y-shaped protein, produced by the body's immune defenses in response to an antigen. An antigen is any bloodstream invader, be it virus, bacterium, or chemical, that the body, through its elaborate surveillance system, perceives as foreign, or "nonself," and makes an antibody against. The antigen-antibody relationship is therefore circular—one cannot be defined without the other.

All antibodies are manufactured by a type of white blood cell, the B-lymphocyte. The arms of the Y act as "claws" to lock onto an antigen target; the lower portion is a demolition unit. When primed by attachment to another protein in the bloodstream, it can blow holes in a cell's walls.

A *monoclonal* antibody is one produced by a clone, or colony, of cells that derive from the same mother cell and so are identical. When an antigen stimulates them to manufacture an antibody, they all make the same one; nothing in nature can copy so exactly.

Before 1975, such pure antibodies were merely an immunologist's dream. To detect the presence of a particular protein inside a cell or in serum, they had to make do with *polyclonal* antibodies—tantamount to trying to catch a single rainbow trout in a pond full of speckled trout. The obvious solution, to breed, hatch, and stock their own "rainbow trout," was denied them because B-lymphocytes, the sole source of antibodies, could not be grown in culture dishes.

What immunologists needed was a practical way to force a cancer cell that can replicate endlessly in a test tube to fuse with a B-lymphocyte containing within its gene repertoire the blueprints for a particular antibody—a "marriage of necessity," in which those cells that fuse survive, and those that don't die off quickly. Such a laboratory creation is called a hybridoma: "hybrid" because it derives from two different types of cells and "oma" because one is a tumor cell. It is literally a cellular factory that can churn out limitless quantities of pure (i.e., monoclonal) antibodies.

The story of the discovery of hybridomas begins in 1974, when Georges Köhler, a postdoctoral fellow working in the laboratory of Cambridge University geneticist César Milstein, got an idea that was either brilliant or naïve and kept him awake much of the night, trying to decide which. For his Ph.D. project at West Germany's University of Freiburg, Köhler had chosen to study the enormous variety of antibodies a mouse's immune system can produce in response to a single foreign invader. The actual number is around a thousand, each attaching to a different site on the amino-acid chain constituting the invader's superstructure.

Köhler's mentor, César Milstein, a scientist of considerable repute, had himself been working for years to solve the riddle of antibody diversity. Although the human genome (the sum total of genetic information) comprises around one hundred thousand genes to direct the manufacture of ten thousand or so essential proteins, the immune system has the capacity to tailor-make a billion or more different antibodies as the need arises. The tantalizing question was: how? Obviously the dictum "one gene, one antibody" failed to account for this incredible diversity. What seemed far more likely was that two or more genes combined in some way to code for the design and production of each individual antibody. To learn more about this process, Milstein had obtained from National Cancer Institute researcher Michael Potter a line of mouse cancer cells that produced a single antibody, though to what no one knew. The challenge was to match the antibody to the precise antigen to which the mouse cell had reacted—the immunological equivalent of the police rounding up a thousand people per city block to hunt down a fugitive whose face was completely unknown to them.

Milstein assigned the task of tracking down the culprit to young Köhler, who got nowhere with it. He couldn't even get Potter's cancer cells to grow in a culture dish. Then an audacious idea struck him: instead of trying to find the needle-in-the-haystack antigen, why not reverse the process: start with a *known* antigen and stimulate B-lymphocytes to make a specific antibody to match it?

It occurred to Köhler to immunize a mouse against a foreign protein, remove the spleen, isolate its lymphocytes, and then fuse them with cancer cells to keep them growing in a test tube. With luck, one of the hybrids so produced might make a pure antibody to the antigen with which the mouse had originally been inoculated. It seemed like a good idea, even a great one, with but one drawback: why had no one ever thought of it before?

The next morning Köhler broached the scheme to Milstein, who immediately saw three problems with it. First, how to get the lymphocytes and cancer cells to fuse in adequate numbers?

Second, how to kill off all the other cells in the suspension so only the fused cells survive? And third, since even a pure antigen could generate a thousand different monoclonal antibodies, how to know you've got the one you want?

In spite of these obstacles, Milstein gave his research assistant permission to try it. For his antigen, Köhler chose sheep red blood cells, since a mouse's immune system usually reacted vigorously against them. For tumor cells, he used a mutant myeloma (bone cancer) strain—a "perfect" cell to use in the fusion, as it turned out, and what made it so was an imperfection. Unlike other bone cancer cells of its type, it lacked an enzyme to protect it against a certain chemotherapeutic poison. In order to survive in the presence of this poison, it had to give up its independent ways and join forces with another cell—in this case, a lymphocyte possessing the missing enzyme. Lymphocytes that failed to fuse with "immortal" cancer cells would die off in the culture dish, as they normally did, Köhler reasoned, leaving only hybridomas after a while. To disrupt the membranes of both types of cells to promote fusion, he introduced a virus into the culture (later changed to polyethylene glycol, an alcohol-derived solvent in popular use as automobile antifreeze). And for the crucial test to determine if any of the hybridomas produced antibodies specific for the sheep red blood cells, Köhler devised a quick screen by first linking some sheep red blood cells to fluorescein, a dye that glows green when its electrons are excited by ultraviolet light. A series of additional steps led to this end point: if any of the pure antibodies locked onto the surface of the sheep cells, the tiny glass well containing them would show a bright green halo.

Shortly before Christmas 1974, with his wife, Claudia, by his side, Köhler ran the final step, and in his own words: "I looked at the first two plates. I saw these halos. That was fantastic! I shouted. I kissed my wife. I was so happy . . . It was the best result I could think of."

Georges Köhler, aged twenty-eight, had achieved the immunologist's dream: pure antibodies.

César Milstein was equally thrilled by the unexpected result, and after many repeat experiments to ensure reproducibility, the two submitted their findings to the British journal *Nature* for publication. Milstein concluded the report by observing, "Such cells can be grown in massive cultures to provide specific antibody. Such cultures could be valuable for medical and industrial use"—perhaps the most memorable understatement since Watson and Crick, who in announcing their landmark discovery of the double-stranded structure of DNA in the same journal years earlier, coyly remarked, "It has not escaped our notice that the specific pairing we have postulated suggests a possible copying mechanism for the genetic material."

It is with such muted trumpets that scientific revolutions are sometimes heralded.

When an editor of *Nature* finally responded to Köhler and Milstein's report, it was with less enthusiasm than they might have expected. In effect, he said to them: *Not bad, worth publishing. But much too long. After all, how many scientists have a pressing need for sheep red blood cell antibodies?*

Milstein shortened the report by half, resubmitted it, and in the August 7, 1975, issue of *Nature* medical scientists learned of the birth of a new technology.

Their immediate response was desultory and disappointing. If, as Milstein and Köhler had ample reason to believe, pure antibodies represented the greatest invention since sliced bread, why, they must have wondered, weren't more of their colleagues popping them into their toaster?

In a very restricted sense, the *Nature* editor had been right: there was no great demand for sheep cell antibodies. Otherwise, he couldn't have been more shortsighted, a myopia shared by officials of the British patent office who completely ignored Milstein's suggestion that the government—his employer—patent the process.

Nevertheless, in the decade to follow, Köhler and Milstein's discovery would shake the scientific world, transforming medical science as much as did X-rays and antibiotics. Their "perfect" cancer cell has reproduced itself quadrillions of times by now and is used in virtually every biomedical facility in the world. With its many uses in both research and clinical medicine, the hybridoma—and its product, the monoclonal antibody—has spawned a multibillion-dollar technology.

A key factor in the acceptance of the hybridoma was the work of scientists at The Wistar Institute, who quickly realized the importance of this new technology and began putting it to good use, producing a barrage of publications. Among them:

- The first report of hybridomas producing monoclonal antibodies against gastrointestinal cancers in 1979.
- The first demonstration that monoclonal antibodies inhibit the growth of certain cancers in experimental animals in 1979.
- The first successful linkage of a monoclonal antibody to a cancer-cell-killing toxin in 1980.
- The first monoclonal antibody to detect the presence of colon cancer in humans through a blood test in 1981.
- The first clinical trials of monoclonal antibody treatment of gastrointestinal cancer, begun in 1979 and reported in 1982.

- The first use of monoclonal antibodies combined with radioactive isotopes to scan the body for the presence of cancer in 1983.

As a diagnostic tool, monoclonal antibodies are so accurate that they can detect the equivalent of a beach ball floating in the Atlantic Ocean. When linked to a visible marker of some sort, a fluorescent dye or radioactive isotope, they can light up cell parts too tiny to be seen by even an electron microscope. This is possible because any organic structure longer than eight to ten amino acid building blocks (a few hundred carbon, hydrogen, oxygen, and nitrogen atoms) can be antigenic—that is, can stimulate the production of an antibody when presented to an immune system lymphocyte in a certain way. This antibody is so precisely constructed that it seeks out and attaches itself only to that amino acid configuration and no other. In a sense, a cell surface is like a forest comprising thousands, possibly millions, of trees, some unique to that type of cell. Monoclonal antibodies not only can detect such disease-related "trees" but can home in on certain structures inside the cell. Along with an even newer technique, DNA probes, they represent the ultimate cellular detectives. In the extraction and purification of substances such as the interferons, present in the blood in only trace amounts, a monoclonal antibody can fish out picogram (one trillionth of a gram) quantities.

Together with recombinant DNA technology, monoclonal antibodies speed scientific progress enormously. Where it once took researchers years and years to isolate, identify, and purify a new hormone or enzyme, they can now do it in months, even weeks. While it may still take them a year or longer to create the right hybridoma to manufacture antibodies to, say, a viral protein, once they do, it can be cloned, frozen, and stored in vats for future use. No one ever has to make that particular hybridoma again.

In 1981, when asked to evaluate and rank 150 emerging technologies, a large group of scientists from the Food and Drug Administration chose monoclonal antibodies as probably "the most useful biomedical discovery for the rest of the century." Their report went on to state that these pure antibodies "may have a bigger impact on medical treatment than anything known to this point."

In 1984, Milstein and Köhler, along with pioneer immunologist Niels Jerne, won the Nobel Prize in medicine and physiology for their monumental discovery. In spite of this, it has been my experience that many well-read and otherwise well-informed people, including some doctors, are only vaguely aware of the enormity of this new technology. One reason is terminology: a newspaper story with the words "hybridoma" and "monoclonal antibody" in its lead paragraph tends to discourage many people from reading on. Even when accompanied by a diagram or two, such reports may fail to convey the sheer elegance of Köhler and Milstein's technique. So before moving on to the many exciting uses of monoclonal antibodies in cancer diagnosis and treatment, let's take a different tack and look at the creation of a hybridoma from a couple of cells' points of view.

BEA was an antibody-producing white blood cell of the B-lymphocyte family and a mini-factory with a single, specific mission in life. When called upon, she manufactured an antibody against a tiny but crucial component of the influenza A virus—a protein that perversely kept changing shape, adding an amino acid here, dropping one there, to elude the vaccine-activated antibodies the immune system unleashed to protect its host from the "flu." To do her job effectively, BEA had to keep dividing, doubling her output of defenders, and for this she required a continuous supply of raw materials. In her original locale, a lake region of the spleen, these needs were easily met. But now, uprooted and torn loose from her neighbors by some mysterious hand, she was adrift in a test tube. With so many cells in so small a space, BEA kept colliding with them. Some were mature cells from the same general location and in the same state of bewilderment as she, which provided some comfort. But others were so immature and alien in appearance that she immediately recognized them as the sort of unruly intruders that occasionally appeared on the outskirts of the spleen, only to be quickly disposed of by its police patrol units. BEA had no idea what was wrong with these cells, why they failed to mature or perform useful work; she had never gotten close enough to one to find out. And for the moment she had far more urgent matters with which to concern herself.

Though she had never been on her own before, BEA knew instinctively that she could not hope to survive more than a few days in such isolation. Without her master, a T-helper cell, close enough by to attach itself to her and issue the proper instructions, any attempt to renew herself through replication would likely be chaotic and fatal. BEA did not dare even to try.

MYELO, a bone-cancer cell adrift in the same vicinity was likewise in deep trouble. His reproductive machinery had broken down and could not possibly be repaired without spare parts from the outside called nucleotides. These usually came in two forms: ready-made pieces such as thymidine or else the chemicals from which they could be made. What he must do—and quickly—

MYELO realized, was find a cell that could supply him with the missing parts, or at least the equipment with which to process them from the chemical debris that, thick as seaweed, floated all around him; to enter into some sort of partnership—even marriage, if absolutely necessary—to get what he needed to stay alive.

MYELO abhorred the thought of giving up his carefree and independent ways to spend eternity laboring away at some monotonous task—and eternity it would be, since he possessed the power to bestow immortality upon his mate. But he had no choice. It was either that or perish.

The lymphocyte he kept bumping into, the one named BEA, had the equipment he needed and MYELO, in turn, the rest of the wherewithal to reproduce. But could this strange creature, seemingly as shy as he was aggressive, be wooed and won? It didn't appear likely.

But on their next collision something unusual occurred; some sea change collapsed their outer walls and literally cemented the two together. By swapping the enzymatic equipment one lacked for the growth instructions the other required, both not only survived the current crisis but went on to produce a new breed of progeny.

BEA soon had reason to regret her hasty decision. The immortality MYELO had promised was actually a vampirelike existence. They were criminals, constantly hunted, unable to live in harmony with other cells. The only good to come out of the alliance was that their offspring made the same antibody BEA did, thus ensuring its continuance.

MYELO was made even more miserable by their forced union. He felt trapped; tied down to a workaholic spouse. Even though the two replicated as frequently as MYELO had before—

roughly every eighteen hours—the fun was gone and all their progeny took after BEA, churning out copy after copy of a product he could not care less about. "Marry in haste, repent in leisure," he thought bitterly as he watched their latest brood of antibodies sail away. Instead of waving, MYELO thrust out an armlike appendage and crossed it with another. Among cells, it was considered an obscene gesture.

And so ends the tale of yet another couple who stayed together for the sake of the children.

3

JIM FEDERLINE was born and grew up in Frederick, Maryland. As the eldest of five children, he enjoyed a close personal and later professional relationship with his father, a plumbing contractor who owned his own company. A good student, eager to emulate his father, Jim was drawn to engineering, and when the time came for college, chose the Drexel Institute of Technology in Philadelphia.

By his own admission, the best thing ever to happen to Jim Federline occurred toward the end of his freshman year at Drexel: he met a slim, attractive female classmate with very blue, very expressive eyes. It didn't surprise me when Tinka confessed that she'd actually asked Jim out on their first date. As I came to learn, she is an exceedingly bright, resourceful, and decisive person—one whom I would be happy to have administer my research program or serve on my hospital's board of managers. Jim and Tinka married shortly after graduating from Drexel and, at the time I met them, had two daughters, ages twelve and fourteen.

Having already decided to join the family business, Jim obtained his master's degree in business administration from

American University in Washington, D.C., and then set out to expand and diversify James Federline, Incorporated, a medium-sized plumbing contracting company whose projects were largely local and modest in scale. Jim succeeded beyond even his own high expectations: from fifteen employees in 1967, the company's work force grew to one hundred over the next decade and ultimately to seven hundred, as it developed heating, air-conditioning, and construction divisions and moved heavily into the sewage-disposal business.

Medical science has long known that the specter of cancer stalks certain families. Whatever the cause, whether they share a genetic defect or an environmental hazard that becomes manifest over time, the Federline family bore such a curse. Its first victim was Jim's mother, who contracted breast cancer in 1960, underwent extensive surgery, and survived. Years later, Jim's only sister developed a premalignant form of the same disease and had both breasts removed. But the worst blow fell in 1972 when Jim's father was stricken and rapidly succumbed to lung cancer at age fifty-four. According to Tinka, Jim seldom expressed his sorrow or sense of loss over his father's death, though the feelings obviously ran deep. One of his younger brothers suffered a nervous breakdown shortly afterward, and Jim himself experienced prolonged bouts of depression, especially after another of his brothers died in an automobile accident. To overcome them, he devoted more and more of his energy to the construction company he now owned, taking it in bold, new directions.

Investing heavily in a novel oxygen-ozone process to stabilize and deodorize industrial sludge and transform it into fertilizer, Jim built a pilot plant in New Jersey that performed so well and generated so many new orders that the firm's fortunes soared. With family tragedy seemingly in abeyance, the next decade proved a happy time for the Federlines. Jim became active in the Young Presidents Organization, an international group of several thousand company leaders under the age of fifty, and was elected local chapter president and regional vice president.

Though at present there is no reliable way to determine when a human cancer begins or how long it remains quiescent before entering an accelerated growth phase, certain rough estimates can be made. Studies in both animal and human subjects have shown that, for a lung or pancreatic cancer to grow from a single malignant cell to a mass of ten billion cells and a weight of around ten grams—the minimal size most cancers must reach before becoming clinically detectable—takes thirty to forty cell population doublings and, with two months the average interval between doublings, at least five to six years.

Bear in mind, though, that this is merely the minimum time. In actuality, it usually takes twice as long, because for most cancers the rate of cell birth exceeds that of cell death by only a small margin. Thus most solid cancers (as opposed to blood-cell malignancies) require ten or more years to grow large enough to produce symptoms. My hunch is that Jim Federline's pancreatic cancer took even longer, and might have begun in 1972–73, when, during his deep depression over the loss of his father and brother, his immune defenses were sufficiently impaired to allow an incipient cancer to take root. Though this is sheer conjecture, the mechanisms by which emotions, via such brain chemicals as ACTH and endorphin, affect the immune system are well on the way to being worked out. Whatever the case, it is a safe assumption that Jim's cancer had been present and feeding off his pancreas for a long time.

Early in 1985, this internal enemy, this betrayal of self, gradually made itself known to him, souring his appetite and sapping his energy. Then came pain—sometimes sharp, stabbing, excruciating, but more often deep-seated and cramping—as the cancer invaded more and more nerve-rich structures. With pain jabbing him awake at frequent intervals, Jim's sleep grew fitful. To get through the day, he had to cope with both physical discomfort and the lassitude he blamed on lack of sleep. Increasingly concerned over how often he was getting out of bed at night to swallow antacids or painkillers, Tinka kept urging him to see his doctor, and finally he did.

To his relief, the extensive series of tests, including X-rays of his entire gastrointestinal tract, that Jim's internist put him through revealed nothing: no ulcer, no gallstones, and most reassuring of all, no cancer. Unable to account for Jim's symptoms but suspecting they might be stress-related, the doctor suggested a lengthy vacation before proceeding with more elaborate investigations, and Jim happily complied, taking his family on a month-long sail aboard his sixty-foot sloop to the Virgin Islands. The trip helped him to unwind, physically and mentally, and for weeks afterward he felt considerably better.

Then came a night, a moment of horrifying self-discovery, that Jim would never forget. Lying in bed, he experienced a cramp in his side, began rubbing it, and for the first time felt a hard, distinct, baseball-sized lump under his left rib cage. The next day his internist felt it, too, and ordered a computerized cross-sectional scan of his abdomen. That same afternoon the doctor summoned Jim to his office: the CAT scan showed a large abdominal mass, most likely a pancreatic tumor, though the lump he could feel was not that but an enlarged, blood-engorged spleen—small consolation, since what had caused the organ to swell was the tumor compressing its venous drainage. That was the doctor's first blast of bad news; after giving Jim a moment to absorb it, and to view the CAT scan for himself, he let loose with the second. The possibility existed that, instead of a tumor, the mass they saw was merely a cyst—only exploratory surgery could determine which for sure—but the internist, a candid man, considered the possibility remote. The likelihood was that Jim had pancreatic cancer; if so, the disease had probably progressed beyond the point where it could be excised by surgery. Palliative measures might slow the cancer's growth, but nothing known to medical science could produce a cure.

Numbly, Jim agreed to exploratory surgery, his thoughts a jumble. As the internist phoned the Washington, D.C., hospital to arrange his admission, Jim's eyes fixed on the CAT scan mounted on the viewbox in front of him and on the tumor, dark gray against a lighter gray background, corroding his core. Unbuttoning his jacket, he felt his left side—as if needing to confirm anew that the horror depicted on the X-ray film was actually inside of him.

In late August, surgeon Luther Grey II, fully prepared to remove Jim's pancreas, spleen, and whatever other nonessential tissues showed disease, if such a cleanout proved feasible, opened his abdomen and, within minutes, closed it up again. With cancer nodules studding his intestines, there was nothing to do but take biopsy specimens for pathological examination.

After the operation, a leading Washington, D.C., oncologist came to see Jim to offer a course of combined chemotherapy and X-ray treatments. Politely but firmly, Jim demurred—the horrible side effects his father had suffered from the same sort of therapy still vivid in his mind—and asked about alternatives.

The oncologist, whose forté was chemotherapy, had no answer. Although significant progress had been made in the treatment of blood-cell malignancies, the leukemias and lymphomas, too little was known about pancreatic cancers, or solid tumors in general, to even come close to a cure. New approaches and treatments were being tried—he named several—but without exception each presented problems: either they were just beginning and exceedingly hard to obtain or they were "randomized and double-blinded" in experimental design—meaning that, if enrolled in such a therapeutic trial, Jim would stand a 50-50 chance of getting a simple sugar solution instead of an active medication. The oncologist promised to make inquiries on Jim's behalf, and he did, but nothing came of them.

"Before the big death come a thousand little ones," Jim's father had told him shortly before his demise, and in the days ahead his meaning became clear. The Federlines never did discover exactly how their friends and neighbors first learned of Jim's disease, but once they did they made the worst of it.

On a sunny afternoon shortly after he came home from the hospital, Jim, Tinka, and their two young daughters, who had not yet been told of their father's illness, were about to go for a sail when a casual female friend, obviously inebriated and overcome by emotion, rushed up to him and wailed, "Oh, Jim, you poor man!" Then, to the Federlines' horror, she blurted, "My mother, my dear, dear mother, died of cancer of the pancreas, too!"

The following evening their next-door neighbor, a plastic surgeon, phoned Tinka to inform her of the morbid and absurd rumor making the rounds: that Jim had terminal cancer and wouldn't live out the week. Not only did Tinka's calm and measured reply render the surgeon speechless, it compelled him to avoid all contact with the Federlines from then on.

Another such incident was so presumptuous and pathetic

as to be comical: straight out of an Evelyn Waugh novel. One evening, as the Federlines were about to sit down to dinner, the front doorbell rang and Tinka answered it. On the doorstep stood their sweet-natured, white-haired neighbor from down the street, a Bible in one hand, a stack of pamphlets in the other, a weapon-sized metal cross dangling from her neck, and a fervid gleam in her eye. Joyously, she announced that she had come to pray with Jim, to prepare him for the exalted moment when he would meet and merge with his eternal soul.

Following that episode and a number of solicitous phone calls, the Federlines left town for a week.

In mid-September, more for Tinka's sake than his own, Jim consulted a renowned oncologist on the staff of the Sloan-Kettering Institute in New York. He did not really expect much to come of the consultation and he was right, though there were surprises—one of which was the cavalier, case-hardened manner of the doctor himself. Their first hint of this came when the famous specialist, returning from a lengthy lunch, strode up to his receptionist and with a groan snatched from her hand a stack of phone messages, all apparently from patients under his care or wishing to be. He riffled through them, crumpled them into a ball, and was about to toss them into the secretary's wastebasket when she said, "You really have to call so-and-so—he's the son of a bank president." Haughtily he replied, "I don't *have* to call anybody!" and swept past her into his private office.

The oncologist was more considerate but scarcely more communicative while examining Jim and reviewing his X-rays. Finally, inviting Tinka to join them in his well-appointed office, he stated unequivocally, "There's no treatment worth a damn for cancer of the pancreas and won't be for years to come. Take my word for it."

Having heard the same pronouncement twice before, though not from so eminent an authority, Jim did not dispute it. But Tinka, her "Irish" up, did, firing question after question.

"I see you know a thing or two about cancer treatments," the oncologist conceded grudgingly. "But in your husband's case, I'm afraid there just aren't any. Oh, maybe some pharmaceutical company somewhere is getting ready to test a new drug or antibody against GI cancer, but if so, I haven't heard of it."

That night, as Jim suffered one of his worst attacks of pain, Tinka, in tears, vowed to do more for her husband than just pray and feed him painkillers. The next day, with the help of Jim's secretary, Sally Menigoz, she began phoning the directors of cancer clinics and research centers throughout the country. But lacking the necessary entrée, she got through to very few of them. To do so took personal contacts—clout—and in an inspired moment Tinka realized what she had been overlooking. Digging out the roster of Jim's Young Presidents Organization, she began phoning anyone on it even remotely connected with the pharmaceutical industry and got results within the week. A vice president of the Rorer Company, who knew Jim personally, took it upon himself to contact The Wistar Institute's Zenon Steplewski.

Although most who use the monoclonal antibody that The Wistar Institute has developed for the treatment of gastrointestinal cancers call it simply 17-1-A, its full name is CO-1083-17-1-A. "CO" is the abbreviation for colon cancer, the tissue from which the antigen, or immunizing agent, was derived; the numbers and letter that follow indicate the hybridoma batch from which the antibodies were originally harvested.

The way one usually makes monoclonal antibodies to combat a certain disease is to inject diseased cells, or a purified portion of them, into a mouse and wait several days for the animal's immune system—in particular, the antibody-producing white blood cells (B-lymphocytes) in its spleen—to react vigorously against the foreign material. The mouse is then killed, its spleen removed, and a suspension of splenic lymphocytes fused with "immortal" mouse cancer cells in the test tube. Since many different hybridomas, each manufacturing its own unique antibody, will be produced in this way, one must then screen them by a method similar to that devised by Georges Köhler in order to isolate and grow in abundance the specific hybridoma one wants.

To serve as an acceptable anti-cancer agent, a monoclonal antibody should, at a minimum, meet three requirements. First, it must seek out and attach itself in high concentrations to a cancer and in very low concentrations, if at all, to the normal tissues around it. Second, its target on the cancer cell should be membrane-bound; that is, fixed to the cell surface and not shed into the bloodstream, where the monoclonal "missile" might make contact with it before reaching its primary target. Third, once the antibody latches on to the cancer cell, it must be capable of triggering its destruction through one immunological mechanism or another—usually by recruiting a multitude of immune system killer cells to pounce on and devour the diseased cell: a complicated process known by the equally complicated name of antibody-dependent, cell-mediated cytotoxicity.

Of the thousands of hybridomas Wistar Institute researchers created and tested against human gastrointestinal cancers, the antibodies produced by batch CO-1083-17-1-A—the violent reaction of a mouse lymphocyte to the presence of colon cancer cells derived from an eighty-three-year-old Texan—best fulfilled these three criteria. And so, on November 25, 1985, my Hurley Medical Center team treated its second pancreatic cancer patient by infusing 400 milligrams of 17-1-A, at a concentration of four quadrillion antibodies per milligram, into his bloodstream.

Jim Federline's treatment went without incident and the following morning he and Tinka flew back to Maryland to spend Thanksgiving day with their children.

A week later, Jim phoned to tell me how well he was doing. He was hurting less and eating more. He had even begun jogging again. Much as these gains delighted me, I was reluctant to attribute them solely to our treatment. They might merely represent wishful thinking on Jim's part—the so-called placebo effect—and if so, they wouldn't last. Realistically, I did not expect them to.

Yet, on January 10, 1986, I received a letter from him that brightened my entire day and made me wonder what manner of magic The Wistar Institute's antibodies might have wrought. Jim wrote: "I've been feeling stronger with each passing day and am working out at the local health club to regain some muscle tone and endurance. Other than an intolerance to some foods (greens) and some nausea, for which medication has been prescribed, I am feeling fairly well . . ."

Because Jim Federline responded so well to his initial monoclonal antibody treatment, I felt a battle had been won, perhaps even a turning point reached, in the ferocious war between his cancer and his immune system. But as time made clear, the conflict had merely entered a new phase, with its outcome far from certain.

4

HOW TO ACCOUNT for The Wistar Institute's fast start and prodigious productivity in the monoclonal antibody field? By general agreement among those individuals most directly involved, the main reason is the atmosphere of the Institute itself.

After decades of dreariness and decay, Philadelphia has transformed itself into an exceptionally attractive city in recent years. Many of its streets and sections now sparkle and pulse with vigor—none more so than the University of Pennsylvania campus on a sunny day. Ivy vines embrace its buildings. Autumn leaves rustle in sidewalk marches. Savory aromas of hot dogs sizzling, chestnuts roasting, and fried rice steaming permeate the air from food vendor carts along Spruce and Walnut streets. Even for those whose college dreams and days are long past, it's a charming spot.

The Wistar Institute, at the corner of Thirty-sixth and Spruce streets, sits on the edge of the campus and maintains close ties with the university, although it is not officially a part of it. While dwarfed in size and in resources by the National Cancer Institute and other well-known research institutions, the Wistar has led advances on many fronts, ranging from viral vaccines to cancer immunotherapy. Part of the explanation for its success lies in the brilliance of its top people and part in the milieu of the Institute itself, particularly the freedoms it offers its faculty. Few of its senior people leave for more prestigious, better-paying jobs with universities or pharmaceutical companies. One reason is that at the Wistar they can work on whatever projects they wish at their own pace; no company edicts or governmental directives tell them what to concentrate on, or when to publish their results. In the words of Dr. "Prem" Reddy, the newest member of the Wistar's senior staff and a molecular biologist who earlier had been offered a professorship at Harvard, "There's more cooperation. It's friendlier. My research will go faster here."

Another reason involves risk taking. As Associate Professor Elaine DeFreitas said in describing the role she played in establishing the link between a human leukemia virus and multiple sclerosis, "It meant taking a chance. It could've turned out to be an artifact, a false lead; after two years of hard work, we could've ended up with nothing. And you know all the controversy our findings have stirred up. If I were a junior faculty member at some university and up for tenure, I might not have pursued it. But here you can take a risk and not be afraid of what will happen if it doesn't work out."

In November 1981, as a part-time medical correspondent for ABC News on the trail of a story, I visited The Wistar Institute for the first time. Before that, being neither a virologist nor an immunologist, I knew only that the Institute bred a famous albino rodent (the Wistar rat) used worldwide for biomedical experiments. I also vaguely recalled reading a medical journal interview with cholesterol metabolism expert, David Kritchevsky, in which he said that the Wistar was a nice place to work. But that was the extent of my knowledge until I became aware that four of the top pioneers in the monoclonal antibody field—Drs. Hilary Koprowski, Zenon Steplewski, Carlo Croce, and Giovanni Rovera—all worked under the Wistar's roof.

I expected to interview them all in a single day, possibly two, should any additional quotes or filming prove necessary, and that would be that. But even with so limited an agenda, I could not help being impressed by the quaintness and contrasts of the Institute: the nearly century-old building alongside a modern, superbly equipped research annex. If the Wistar were a cell, I mused, it would be a hybrid, a fusion of the young with the old.

The main entrance to the Institute is through the original building, a three-story sandstone-and-brick structure at the south end of the promenade that cuts across the University of Pennsylvania campus. Above an archway its name is lettered in concrete and there is a frieze of two Greek-robed women, one holding an open book and the other a frond. Worn marble stairs lead to a large lobby with a museum to the left, an auditorium to the right, and a security desk and wrought-iron staircase straight ahead. On a bright day, the lobby is illuminated by sunlight streaming in through a huge skylight to reflect off the black marble floor.

To the rear of the lobby hangs a photograph, magnified thousands of times and blown up to the length of a broom handle, of a great Wistar Institute contributor: the WI-38 cell, the progenitor of one of the first human cell lines that would grow well in a culture dish and so could be used to hatch viruses.

Once occupying most of the building, the Wistar's anatomical museum is now confined to a tennis court-sized space on the first floor. It is replete with preserved specimens of human embryos of various ages and fetal skeletons showing striking developmental abnormalities. Two in particular—a one-eyed cyclops and a two-headed infant monstrosity—are as bizarre a pair of oddities as I have ever seen. Groups of schoolchildren visit the museum from time to time, but otherwise it stands empty, its large area coveted by Wistar scientists cramped for lack of laboratory space.

On my first tour of the Institute, I found the transition from sunlit lobby and museum to fluorescent-lighted, equipment-packed research labs abrupt and jarring, and there were more surprises to come, such as the small, elegantly furnished dining rooms donated by French and Polish benefactors. The most memorable experience, however, was my introduction to The Wistar Institute's longtime director.

By any definition, Hilary Koprowski (H.K. to some of his younger faculty) is an extraordinary human being—as comfortable at the keyboard of a piano as in a virology laboratory and a virtuoso of both: a researcher and teacher who, for over thirty years, has inspired a succession of disciples to push forward the frontiers of science. Physically compact and looking much younger than his seventy years, he has a penetrating gaze, a Middle European accent pleasing to the ear, and seemingly effortless charm. He is, in fact, European to the core: multilingual and extremely knowledgeable about the arts, a raconteur and bon vivant, imperious or playful as the mood strikes him. Depending on his schedule, he can be seen at the Institute in conservative business suit and tie, sports shirt and Bermuda shorts, or designer coveralls. He is a keeper of promises, large and small. On the basis of personal experience, it's my guess that he has earned the friendship and devotion of more people through the little attentions he pays them than through the big favors his position enables him to bestow.

Just about everyone who knows its workings agrees that Hilary Koprowski embodies the spirit of The Wistar Institute. A current of "what next?" excitement flows from the director's second-floor office to pervade the entire staff. Carlo Croce refers to the narrow corridor outside Koprowski's office as the "Khyber Pass" of the Institute, as well as one of the most dangerous spots in the scientific world. Should Koprowski emerge from any one of three doors just as one of his younger scientists saunters by and should he begin a conversation with the dreaded words "Oh, by the way . . ." that hapless individual might find himself sent on a mission to a Swiss laboratory, a French pharmaceutical company, or a Rome museum. Though the freedom to refuse such requests is implicit, it is seldom exercised.

I have heard Hilary Koprowski's thirty-year reign as Wistar Institute director described as a monarchy, an oligarchy, a benevolent dictatorship, even as a medieval fiefdom with Koprowski as lord—though almost always with a good-natured grin and no wish whatsoever for it to change. If anything, there is understandable concern over what might follow the inevitable end of the Koprowski regime and the cohesive force he himself represents. Yet, had it not been for a chance encounter with an old friend on a Rio de Janeiro street in 1940, Koprowski and The Wistar Institute might never have come together at all.

Born in Warsaw in the middle of the First World War and destined to be permanently uprooted from his Polish homeland by the Second, Koprowski spent much of his youth vacillating between a musical and a medical career. His dream was to become a concert pianist, and from the age of five he strove diligently to achieve it. But ultimately the practical side of his nature convinced him that medicine offered the more secure future—though even this seemed in doubt when the Second World War erupted and, with his new bride, Irena, herself a physician, Hilary fled Poland for Italy, Spain, Portugal, and eventually Brazil.

Arriving in Rio de Janeiro in 1940 with his wife and infant son, Hilary made no attempt to obtain a Brazilian license to practice medicine, knowing how long it would likely take, but instead tried to supplement the little money he had by teaching piano to children. Then one day, racked by uncertainty over what to do with himself, he set out on what would prove to be the most eventful stroll of his life. Along the way he passed a man who immediately cried, "I know you!" Almost unable to believe his eyes, Hilary stared back at an old Warsaw high school classmate and exclaimed, "I know you, too!"

Quickly, Koprowski's friend, a fellow physician, helped him get a job doing yellow fever research for the local branch of the Rockefeller Foundation, and within a dozen years Hilary had won recognition as one of the world's premier virologists, a Nobel Prize candidate for his successful oral polio vaccine. His spectacular success and The Wistar Institute's reentry into the mainstream of modern science would eventually coincide.

In 1892, to commemorate his great-uncle, Dr. Caspar Wistar, Philadelphia lawyer Isaac Wistar purchased land from the University of Pennsylvania upon which to build an institute where "anyone in search of new and original knowledge would be welcome to work." Caspar Wistar was a post-Revolutionary War surgeon, the first medical school professor of anatomy in the United States, an intimate friend of Benjamin Franklin's, and a dedicated physician who risked his life many times fighting the yellow fever plague that struck Philadelphia in the early 1800s.

Before becoming a prosperous lawyer, Isaac Wistar was a bold and sometimes unlucky adventurer: an Indian fighter, cattle trader, gold seeker in California, and Union general who permanently lost the use of his right arm in the Civil War battle of Ball's Bluff. Isaac had a twofold purpose in founding The Wistar Institute of Anatomy and Biology: to house his Uncle Caspar's extensive collection of anatomical specimens and to provide a suitable workplace for anyone wishing to pursue scientific knowledge "in a spirit of free inquiry." It came into existence at a time when it was sorely needed: a period in our medical history when those unfortunate enough to fall ill were still being bled, a pint at a time, of "foul blood," purged of bowel poisons, or subjected to even worse tortures. Physicians in general were held in such low esteem that a medical student was reputed to be "the only son of a family who was too lazy to farm, too stupid for the bar, or too immoral for the pulpit."

Some men *say* they'd give their right arm for a cause, but Isaac Wistar actually *did* it. He bequeathed his nerve-dead arm to the Institute for whatever might be learned from studying it. But since no one on the staff has ever had the heart to dissect it, the arm, residing in an airtight container in the main safe, remains intact.

That Isaac Wistar had a sense of humor, as well as a sense of destiny, is illustrated by another of his bequests to the Institute—in anticipation of a celebration that, if it took place at all, would be ninety-one years in the future:

> This bottle of Jamaica rum was left in the house of Lydia Cooper in Arch Street near the Delaware River by some young British

> officers who were billeted upon her, when they were suddenly ordered to their commands on the evacuation of Philadelphia by Sir Henry Clinton on the 17th of June, 1778. . . . This bottle has only been opened twice since. In 1850 a spoonful was taken from it for a dying invalid, but not agreeing with her it was again sealed up. It came to me about 1894 through my uncle, Franklin C. Jones, when I opened, tasted, and recorked it. I now give it to the Wistar Institute of Anatomy and Biology with the request that it be retained till A.D. 1992 and then drunk by its Board of Managers at a dinner to be arranged by them for that purpose at the Institute's expense on the completion of the first century of its corporate existence. . . .
>
> Isaac Jones Wistar
> January 1, 1901

Among distillation experts at the Institute, a difference of opinion exists about the likely quality of the rum. "But," David Kritchevsky is quick to add, "you have to admit the old boy sure knew how to stretch a drink."

For many years The Wistar Institute functioned mainly to preserve and display Caspar Wistar's collection of anatomical specimens and add to it as opportunities arose. Its small staff also undertook studies in neurology, comparative anatomy, and eventually embryology—out of which came the purebred and still widely used Wistar rat. But with the advent of antibiotics and other major medical advances, anatomy was dethroned as "queen of medical sciences," general interest in related fields of study waned, and the Institute lapsed into a prolonged period of near-total inactivity. Which is not to say it was abandoned entirely. It still had its museum, its printing press for the publication of anatomical journals, and its esoteric collection of specimens.

Once, when pioneer Canadian neurosurgeon Wilder Penfield was asked to lecture at the Wistar, he accepted the invitation with the proviso that, instead of a lecture fee, he be given the brain of Sir William Osler—the most famous of modern-day physicians—which the Institute just happened to have on hand.

In the mid-1950s the Wistar board of managers finally recognized how passé the work of the Institute had become and took the advice of University of Pennsylvania Medical School Professor Geoffrey Rake, whose lab was then housed on its premises, to hire a well-known virologist as Institute director. Hilary Koprowski, Lederle Laboratories assistant director in charge of virology research, president of the New York Academy of Sciences, and creator of the first successfully tested oral polio vaccine, was an obvious choice.

In the spring of 1957, Koprowski paid his first visit to The Wistar Institute, beholding with bemusement the massive skeleton of a whale that filled most of its third floor and the bone-packed rooms in its basement, but taking more careful notice of the building's spaciousness than the oddities in it.

Returning to Pearl River, New York, he discussed the job offer with his wife, Irena, his Lederle Laboratories friend and colleague David Kritchevsky, and several other young scientists he knew there. When they all agreed to accompany the Koprowskis to the Wistar and help transform what was little more than an empty shell into a modern research institution, he accepted the directorship. His first executive decision was to dispose of the whale skeleton, even though it would take four freight cars to transport it to its new home, the Field Museum of Natural History in Chicago.

Koprowski's next accomplishment was to recruit a group of bright young scientists to work at the Institute—one of whom was West Germany's Eberhard Wecker. The visiting researcher slept in a makeshift bedroom on the third floor, next to which was a bathroom with shower. One sweltering summer night, Wecker, the sole live-in scientist, stepped naked from his room to head for the shower, only to hear a gust of wind slam shut the bedroom door. To his horror Wecker realized that he was both naked and locked out. His only hope was to intercept the night watchman on his midnight rounds and have him use his master key to reopen the bedroom door. Sweating and shivering at the same time, Wecker descended the stairs to the darkened museum and began calling out for the watchman, his sense of embarrassment so acute that, even to his own ears, his voice sounded hollow and high-pitched.

Drawn by the strange and totally unexpected cries, the young guard entered the museum from the far end, saw a pale, naked, arm-flailing apparition rushing toward him, and promptly spun around and fled, convinced that his worst fears about the job he'd just begun were justified: amid all the dry, misshapen bones lurked a ghost! Screaming, the guard ran for his life. Wecker, startled and mortified, but feeling that he had no choice, ran after him, down the back stairs and along the winding corridors of the basement until, coming upon the guard fumbling with his keys to unlock an outer door, he finally caught up with him.

Eberhard Wecker completed his studies at the Wistar and returned to Würzburg, West Germany, where he currently directs its Institute for Virology and Immunobiology. The twist to the story turns not on him, but on the night watchman, who went on to medical school and is now chief of neurosurgery of a large Florida hospital.

The international awards and honorary degrees Koprowski has garnered over the years number in the dozens. Yet it would be presumptuous, if not foolish, to say he has reached the pinnacle of his career. In the words of Wistar Associate Professor Meenhard Herlyn, "Hilary has the long breath." A researcher, like a diver for sunken treasure, must sometimes plunge into deep, dark waters and stay under until either his search is rewarded or he runs out air—i.e., funds, faith, or fresh ideas. With his agile brain and in his capacity as Institute director, Koprowski can pursue an elusive concept longer than most, and on several occasions this "long breath" has proved invaluable to him.

One of these involved an experiment that he fondly refers to as "an extraordinarily successful failure." In 1961, Wistar investigators, on their thirty-eighth try, found a strain of normal human cells that would grow luxuriantly in a test tube—a finding of such immense importance and usefulness, as it turned out, that it formed the basis for the studies of a full generation of Wistar scientists.

Up until then the only human cells that would grow for long periods in tissue culture were cancerous in origin, which led to the theory that most cancers develop when tumor-producing viruses, latent in human cells, are activated. To test this hypothesis, normal human cells whose growth could be sustained long enough to observe the possible emergence of these alleged cancer-causing agents were needed, and so the National Cancer Institute contracted with The Wistar Institute to perform such studies on its WI-38 cells.

The team assigned to the project never did spot any tumor viruses hatching from their cells, but just about any other virus introduced from the outside grew plentifully and could, by various laboratory manipulations, be transformed into a vaccine.

Around this same time, two events made the manufacture of new and safer viral vaccines a pressing public health issue. Five million Americans were unwittingly inoculated with a polio preventative derived from monkey cells later found to be contaminated with SV-40, a virus known to cause cancers in animals. And in the mid-1960s, an epidemic of German measles (rubella) infected so many pregnant women and produced so many fetal abnormalities that development of a vaccine against this disease became a public health priority.

By 1968, the Wistar group, led by Drs. Stanley Plotkin and Koprowski, had the proof that not only polio but also rubella and rabies vaccines could be produced faster and in a safer form using human rather than animal cell lines. Announcing their discovery to the scientific world, they geared up to conquer rubella—only to find that the technical problems they had solved were trivial compared to the political obstacles they faced.

Despite extensive studies that the Wistar and other groups had carried out to refute the latent-virus theory of cancer causation, Food and Drug Administration officials still clung to the notion that human cells *might* contain such tumor viruses and therefore only animal cells should be used to manufacture vaccines.

Hilary Koprowski spent years trying to convince the FDA and its scientific advisers that, in fact, the opposite was true; moreover, that vaccines derived from human sources were preferable to those from animals because they were far less likely to provoke allergic reactions. But his efforts were in vain. So, as he had done once before to establish the potency of his oral polio vaccine, Koprowski persuaded a European pharmaceutical company to conduct clinical trials of The Wistar Institute's rubella vaccine. These proved so successful that the vaccine is now the preferred method of preventing rubella throughout the world.

Particularly in the viral vaccine field, the discoveries of Wistar scientists have, without exaggeration, saved millions of lives. I have never actually seen a case of rabies; I pray I never will. I have, however, viewed with mounting horror a film of a young African in the terminal throes of this disease: his back arched like a bow, his limbs thrashing and jerking convulsively, the copious amounts of saliva produced by the viral infection causing him to literally foam at the mouth. Unlike smallpox, recently eradicated from this planet, rabies finds ample prey among the animal kingdom and a convenient means of transportation from one victim to the next in their saliva. Yet, thanks to the safer, easier-to-administer, and more effective vaccine The Wistar Institute has made available, doctors seldom have to worry about rabies developing among the countless patients they treat for animal bites. Nor are they likely to see very many more children suffering from the brain inflammations or birth defects that occasionally follow rubella infections.

A more recent example of Koprowski's "long breath" involves The Wistar Institute's commitment to monoclonal antibody research.

Between 1955 and 1975, National Cancer Institute scientists tested approximately 400,000 different compounds for their anticancer properties. Some were naturally occurring substances from the sea, soil, plant life, or fungi, but most were specially synthesized for this purpose. The yield from this extensive and expensive screening program was deeply disappointing, only eight of the 400,000 substances proving safe and effective enough for human use.

With some exceptions, the chemotherapy of *solid* tumors—in contrast to that of the blood-cell malignancies, the lymphomas and leukemias—has been and will likely continue to be a failure. Just as the saturation-bombing tactic employed toward the end of the Second World War usually demolished not only its military target but also everything around it, so too with solid-tumor chemotherapeutic regimens; their destruction of cells is too indiscriminate. A more precise, guided missile or "smart bomb" attack was clearly needed, and the creation of hybridomas and monoclonal antibodies made it feasible.

Many cancer experts saw this, wrote editorials in medical journals about it, but proceeded cautiously. So did Wistar Institute scientists, even though related research done previously put them a step or two ahead of the field.

In 1975, Walter Gerhard, a Swiss-born physician and recent addition to the Wistar faculty, was one of the first immunologists to isolate pure, or monoclonal, antibodies from the serum of mice immunized against influenza virus. But Gerhard did it by a very cumbersome method and was still looking for an easier way when Köhler and Milstein published their hybridoma paper. Other Wistar scientists interested in the new technique included geneticist Carlo Croce, an expert in fusing mouse with human cells to determine the chromosomal location of certain genes; Zenon Steplewski, who had long been working on cell fusion and cancer-cell preservation techniques; Barbara Knowles, intent on learning more about how the mouse immune system dealt with cancer; and Dorothee and Meenhard Herlyn, a husband-and-wife team from West Germany, who, in addition to being veterinarians, were skilled laboratory investigators.

In 1977, by the exact same process depicted earlier in the story of BEA and MYELO, Koprowski, Croce, and Gerhard produced monoclonal antibodies to a portion of the influenza A virus—thereby satisfying themselves that the modifications they had made in the basic hybridoma technique worked. They then produced and screened a variety of monoclonal antibodies against human tumors. One of these, the 17-1-A antibody, proved fairly selective in targeting colon cancer cells for immune system destruction and so was subjected to further in vitro and in vivo studies.

When injected into an assortment of tumor-bearing mice and other experimental animals to test both its anti-cancer effects and its safety, the 17-1-A antibody not only performed well, it exceeded all expectations. Further experiments by Zenon Steplewski's team showed that one way these antibodies exert their tumor-destroying effects is by recruiting immune system killer cells called macrophages to help in the attack. Since macrophages are known to have receptors, or slots, for the bottom portion of the Y-shaped antibody molecule, one can visualize its top portion locking onto a cancer cell and bringing it and a macrophage together—though this is probably an oversimplification of a more complicated process. Nonetheless, if before 17-1-A antibody treatment tumor-bearing mice are given injections of silica, a substance that inhibits many macrocyte functions, the tumor-killing actions of the antibodies are largely abolished.

Hilary Koprowski viewed these experimental results with something more than strict scientific interest. His mother, on no less than four separate occasions in her life, had been stricken by cancers of the ovary, throat, and both breasts. Each time, as she underwent therapy of one type or another, Hilary wished he knew enough about the nature of these mysterious maladies so that he might be able to contribute to their conquest. But even with so powerful a new weapon as monoclonal antibodies, he had to wonder how much a comparatively small institute, such as his, could do to defeat such a scourge.

For at least one of its victims, a South Carolina businessman whose story Jim Federline knew and took heart from, the answer was: a great deal.

5

IN LATE NOVEMBER 1982, Norman Arnold, a prosperous merchant and civic leader, his wife, Gerry Sue, and their three sons flew from their home in Columbia, South Carolina, to a small island in the Bahamas. It was the day before Thanksgiving, the kids were out of school for the rest of the week, and the Arnold family was determined—grimly so—to have a good time, though their jaws must have ached from so many forced smiles. In truth, it was the low point of Norman Arnold's life: a nightmare growing more and more grotesque from which there seemed no wakening. At age fifty-two, Norman was dying and had planned this trip not so much to relax as to say goodbye to his teenage sons in his own way.

Five months earlier, Norman had been fit and athletic: five foot seven and a trim 160 pounds. Now he was frail and balding and weighed only 112 pounds. In his own words, he "looked like a reject from a Nazi death camp." Yet he was in better spirits these days than a month before when a well-meaning friend, who belonged to the same synagogue as the Arnolds, had suggested that maybe, just maybe, they might want to advance the bar mitzvahs of their twin sons by several months.

Norman describes himself as an intense "take charge" personality always on the go. Even four months after being told—repeatedly—that he would probably not live to see 1983, he had not appreciably slackened his pace of living. He had, in fact, zipped through the five stages of the dying process, from denial to acceptance, in nearly record time.

Although his time of crisis preceded Jim Federline's by a few years, there are many parallels in their stories. Like Jim, Norman never suspected he was harboring a pancreatic cancer; it is such an insidious disease that few of its victims do. Yet, throughout the spring of 1982, he had sensed something wrong. He suffered a nagging pain in his right flank—probably a pulled muscle from one of his vigorous tennis matches. He felt nauseated frequently—too many cigarettes. He was listless and depressed—from all his business worries, no doubt. The thought that he might be seriously ill crossed his mind, but it didn't figure to be cancer.

In 1962, Norman's father, Ben, who had started the wine and beer distributorship he now ran, died suddenly of a heart attack at the age of sixty-two. From previous medical examinations, Norman knew he had a high serum cholesterol level and the tendency to develop diabetes. But despite this, he continued to smoke cigarettes, push himself hard, and consume half a gallon of chocolate ice cream several times a week. If any ailment was going to kill him, it looked to be coronary artery disease.

His good friend and tennis partner, Richard Helman, a gastroenterologist, kept after Norman to get a checkup until finally he did. The test results showed he had gallstones—not a malady anyone would want, but not usually life-threatening either and a likely explanation for his symptoms.

"Get your gallbladder out!" Helman urged, but Norman kept procrastinating. It was summer; his tennis game needed work. Helman did better with a more compelling argument: "Do it now, before cholesterol and cigarettes close off your arteries."

On July 28, Norman underwent surgery at a local hospital. Upon entering his abdominal cavity, surgeon Dan Davis, another longtime friend, found not only the diseased gallbladder he was expecting but something completely unexpected: an egg-sized tumor in the pancreas and, worse yet, a dime-sized metastatic cancer deposit in the liver. In despair, Davis went ahead and removed the gallbladder, wishing he could take the cancerous pancreas with it, but because the disease had already spread to the liver, it was beyond his power to cure.

While his patient slept in the recovery suite, Davis trudged to the surgical waiting room to break the bad news to Norman's wife and mother. Gerry Sue was so stunned by the enormity of Davis's disclosure that she couldn't comprehend it at first, making him repeat his findings over and over. When their meaning finally sank in, she pleaded with Davis not to tell Norman right away, to give him a week's grace to recover from the surgery. But much as Davis might have wanted to, he did not feel it would be ethical.

Groggily, Norman Arnold listened to his surgeon friend—a man he had played horseshoes with and joked with at a Fourth of July party just three weeks before—as he told him he had inoperable pancreatic cancer and the standard four to eight months to live. Even after a parade of consultants had confirmed the diagnosis, and the death sentence, Norman still could not bring himself to believe it. He tried denial, withdrawing into himself, burying his head in a pillow for the next few days. Then, moving on to the next stage in the dying process, he got angry, tired of being pushed and poked and treated as if he were a helpless, half-dead invalid. He could stand the pain but not the pity. Old friends who came to visit seemed not to know what to say to him and to be in too much of a hurry to leave. He felt like a leper. In the eyes of one lugubrious relative, he was already a corpse; Norman asked her not to come again.

On the seventh postoperative day, he reached his limit. Neither his wife nor his surgeon could dissuade him; he checked out of the hospital. Davis had to come to Arnold's home to take out his stitches.

Norman had served four years as a U.S. Naval officer in the Pacific during the Korean War. He had faced perilous situations before. But this was different; this enemy had no face, no known objectives, and attacked from within. His doctors had been candid; brutally put, there existed three ways to combat internal malignancies—"cut, burn, or poison"—but none worked well, or for long, with pancreatic cancer. Even so, Gerry Sue kept reminding him, he didn't have to battle on alone. Long active in civic affairs, Norman had served as chairman of the Children's Bureau of South Carolina, the Governor's Economic Task Force, and the Richland County Heart Fund. He had even founded the local chapter of the Boys Club of America and started the Columbia Zoo. Now came the payback. With his many friends rallying around, Norman decided to make a fight of it. He read everything he could find on cancer research. He made the rounds of the big medical centers. His cousin, Charleston physician Charles Banov, heard about the therapeutic trial Wistar Institute investigators were conducting with monoclonal antibodies and looked into it, but was left with the impression that they weren't ready to treat patients other than those with colon cancer.

Interferon was still being touted as an all-purpose cancer killer, but was in short supply and hard to come by. A friend of Norman's in Toronto contacted an oncologist in the United States who offered a two-week course of interferon therapy for a fifty-thousand-dollar fee, payable in advance. Closer to home, an immunologist at the Medical University of South Carolina agreed to try to make a monoclonal antibody to attack Norman's cancer, if he provided the money to purchase the necessary equipment.

Norman did help finance the purchase of the equipment, but the process was agonizingly slow, and he soon came to the conclusion that he didn't have time to wait. His weight was dropping alarmingly; he was growing weaker and bleaker. Something had to be done now, if only to buy a little extra time.

In August, Norman received his first chemotherapy treatment at the Georgetown University hospital. He underwent two more courses of treatment at a local hospital before deciding to forgo them in favor of something else. As he told a reporter sometime later, "The chemotherapy was draining me so much, my weakness was worse than my pain. Once, I lost my balance, fell, and had to have a dozen stitches in my head. It was making my life intolerable. I simply didn't want my boys to remember me this way—as a cripple. If I had to die, at least I'd die in some more dignified manner."

In the meantime, Norman had read about an unconventional treatment method, the "macrobiotic diet," and began following it. The diet consisted mainly of whole grains, such as rice, oats, wheat, and corn; no meat, dairy products, or sugar. Norman didn't expect miracles, but at least it was something he could do for himself, and he needed that.

His doctor friends belittled the alleged healing power of the diet and convinced him that something more should be tried. So, in September, Norman flew to Philadelphia to meet with Hilary Koprowski and Dr. Henry Sears, the bold, young surgeon from the Fox Chase Cancer Center who had taken the lead in testing the effectiveness of the Wistar's 17-1-A antibody in a small number of cancer patients.

Norman liked the concept of marshaling the body's own defenses through use of the antibodies; it made good sense. Also, the treatment had no apparent side effects. But there were imponderables as well: Sears had treated only a handful of patients, all with colon cancer, and they had been either too riddled by disease to begin with to hope for a favorable response or treated so recently that no conclusions could be drawn. Moreover, Henry Sears was undecided about including pancreatic cancer patients in this pilot study.

Sears promised to discuss his case further with Hilary Koprowski and let Norman know. Two weeks later, Norman underwent his third course of chemotherapy and found it as debilitating as before. He had lost forty pounds, all of his body hair, and most of his scalp hair. He looked, and felt, decrepit, years older than his age. He phoned Henry Sears: had they reached a decision in his case? They had; they'd give the treatment a try.

In late September, Norman spent three days at the American Oncological Hospital in Philadelphia and was infused with 430 milligrams of 17-1-A antibody. The treatment itself went well, but when he asked Sears whether he should continue with the chemotherapy, the answer was yes; no one yet had been cured of cancer by monoclonal antibodies.

In October, Norman received two more courses of the drug 5-fluoruracil and, feeling worse than ever, decided that was enough. He would adhere to the macrobiotic diet, practice the "guided mental imagery" techniques he'd learned in Dallas, but no more chemotherapy or frantic search for a miracle cure; his quest had become so all-consuming that he was neglecting his three young, very frightened sons.

So, for Thanksgiving vacation, he took his family to Abaco, the small island in the Bahamas that he and Gerry Sue had visited before. The weather was balmy, the waters sparkling, and the accommodations as good as he remembered them. Through sheer force of will, Norman managed to overcome his acute sense of impending loss, of "last looks," of time passing faster than ever before.

He and his sons talked quietly and at length. Though he never came right out and said he expected to die soon, they could see for themselves how sickly-looking he was at 112 pounds. It was a little thing, but he was particularly proud of them when, knowing about his special diet, they ordered fish, not turkey, for Thanksgiving dinner. With Gerry Sue's support, he had done what he set out to do in meeting his obligation to his sons. Was that what was making him feel so much better?

For as long as he could, Norman resisted falling into the mental trap of thinking that one of the treatments he had received was finally beginning to work. But as the days sped by on Abaco, and later at home, he gradually gained weight and strength and there was no denying he *did* feel better. Was the improvement real, Norman wondered, or "all in his head"? Finally, at Gerry Sue's insistence, he saw his doctor and underwent further tests. An ultrasound study of his abdomen in early January and another in February showed that his tumor masses were shrinking steadily. His hair grew back. His main problem now was how to get his friends and relatives to notice that he was no longer dying.

His doctors were intrigued, but urged caution; he was not yet cured, merely in remission. Norman's answer was to whip as many of them in tennis as would play him. Throughout this period, he not only stuck to his macrobiotic diet but promoted it to his friends, even the doctors among them, who remained more impressed by his tennis game than by what fueled it.

In June 1983, Norman's physician gave him the best possible news: his latest CAT scan and ultrasound examinations were normal; all traces of the tumor gone. Repeat studies in November 1983 and September 1984 were also negative, and as of this writing, Norman Arnold remains symptom-free.

Most doctors who treat cancer patients, myself included, use the word "cure" sparingly. Some superstitiously avoid it altogether; others, when its use seems justified, speak it in humble tones. So, to hedge, let's just say that Norman Arnold represents "a five-year cure." What did it? Norman is convinced his "large life" diet played a significant, possibly a dominant role; the opinion rendered by the American Cancer Society that "no evidence exists to show that macrobiotic diets result in objective benefit in the treatment of cancer in human beings" does not dissuade him.

Like Norman, I have no clear concept of what worked in his case; I only have prejudices. As a physician, I know that an adequate nutritional intake is needed for one's immune system to function optimally; that malnourished patients undergoing emergency surgery suffer up to ten times the rate of postoperative infections than do well-nourished ones. I also remember the empiric way in which the medical profession was once forced to treat tuberculosis sufferers. Though cures were uncommon, plenty of rest, good food, and sunshine *did* prolong the lives of those fortunate enough to receive these supportive measures. Then along came the anti-tuberculosis antibiotics, streptomycin and isoniazid, and although adequate amounts of rest and nourishment were still needed, antibiotic-treated patients, with few exceptions, were actually cured. In brief, specific treatment, when available, is better than nonspecific; but I won't belabor the point.

In a letter to me, Norman wrote, "It is probable that both the monoclonal antibodies and the macrobiotic diet contributed to my recovery," and for the moment I have no solid evidence with which to challenge this sincerely held belief. But in my own biased view, it was mainly the monoclonal antibodies that triggered the healing process. I hoped they would do the same for Jim Federline.

6

THE FIRST few minutes of a monoclonal antibody infusion are tense; an allergic reaction to its mouse protein component is always possible and, in its most extreme form—anaphylaxis—life-threatening. To safeguard against it, we routinely skin-test the patient with a minute quantity of the 17-1-A antibody first and have a full complement of emergency equipment and drugs, such as Adrenalin and cortisone, handy. No patient yet has suffered any adverse reaction, allergic or otherwise, from his first antibody infusion, but with re-treatment the potential risks—and my anxiety level—rise steeply. If nothing bad happens with the first few cubic centimeters, nothing is likely to later, and all involved in the procedure can begin to relax.

It's at this point that my mind sometimes wanders: I watch the slow, steady drip of antibody-coated white blood cells into the patient's vein and try to visualize where they go, what they do, after entering the miles of capillaries, lymphatic channels, and extracellular spaces in the body. As AIDS victims have tragically shown, without a healthy immune system to protect us against the countless microorganisms and chemicals that enter our bloodstream from the outside daily, we are all helpless. We are likewise imperiled from within.

Let's say, for argument's sake, that the odds of one of our cells turning cancerous are a trillion to one. Or better yet, ten trillion to one. They might seem like long odds, but since our grown body contains around a *hundred trillion cells,* the probability is that it has to deal with up to ten new cancer cells every day. It is a tribute to our immune defenses that two-thirds of us will never develop clinically detectable cancer in our lifetimes.

Before going into detail about the workings of our immune systems, I would like to take up two issues pertaining to proportionality. First, the matter of size and space. On first thought, cell dwellers—the billions of proteins and particles within the confines of a single human cell—might seem tightly packed together, but actually they are more like ranchers than big-city folk. They have plenty of room in which to move around. The average white blood cell, for example, is around fifteen micrometers in length and one billionth of a gram in weight. A sheet of five million such cells could fit into one square centimeter—less than the size of a fingernail. But in the bloodstream there are usually only five to ten thousand of them per *cubic* milliliter—so no crowding there. A protein antibody is approximately four million times smaller than the cell producing it and comprises hundreds of amino acids. So, even though an individual cell is orders of magnitude smaller than we humans, proportionally its inhabitants enjoy as much space for themselves as we do here on earth.

Now for a more metaphysical matter. If one defines intelligence as "the ability to choose successfully between alternatives," then cells do demonstrate something of the sort. They communicate with one another, they collect into networks, they

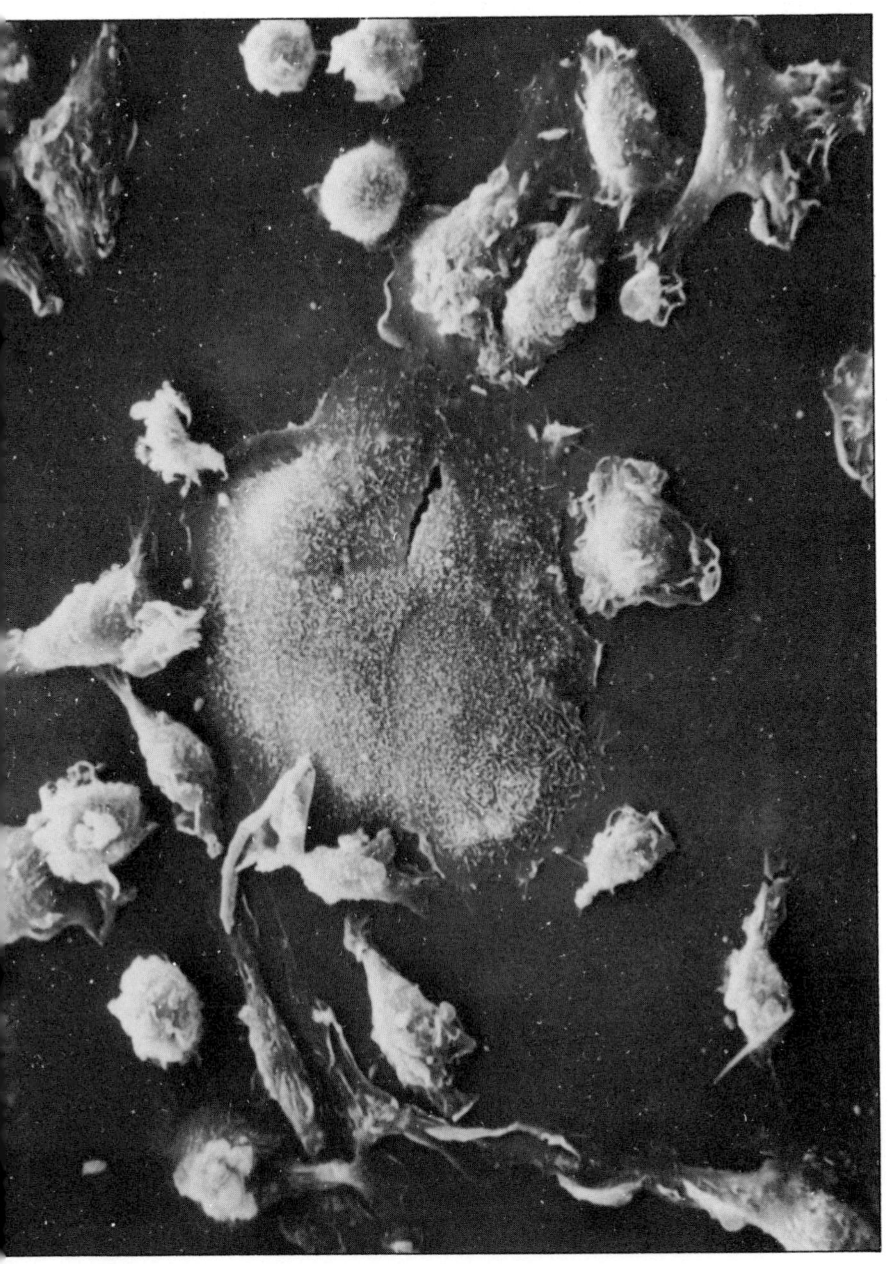

Electron photomicrograph of macrophages attacking a cancer cell. Photo by Professor Zenon Steplewski of The Wistar Institute

perform specific tasks. But so do the microcircuits of a computer and in an analogous, on-off fashion. The more intriguing question is: are cells "conscious," in possession of even the most rudimentary sense of their place in the order of things? To get to the point: can they think? In the case of T-helper cells, the principal decision makers of the immune system, my hunch is that they can, at least insofar as determining what is "self" or "nonself," harmless or harmful to the body. It is upon the accuracy of such decisions that our health depends.

Approximately 1 percent of our body weight consists of white blood cells, called lymphocytes because they travel through lymph channels and collect in lymph nodes. On the average a trillion strong, this cell force functions largely, though not entirely, independently of the brain and is not hard-wired to it, as is the eye or the ear. Its members do, however, communicate with various brain centers through chemical signals called lymphokines and monokines, and neuroimmunologists have yet to learn which system, nervous or immune, dominates when it comes to combating certain diseases; which literally "calls the shots."

Evolving and refining itself over eons, the immune system consists of three major divisions, each with several subdivisions, to carry out its two basic functions: intelligence gathering and defense. T-cells (so designated because, though they develop in the bone marrow, they are "educated" in the thymus gland) comprise the bulk of the immune system forces and serve mainly, though not exclusively, to regulate the function of the second category, the B-lymphocytes; T-helper cells stimulate their antibody production and T-suppressor cells limit it. The third major division consists of hunter-killer units, such as macrophages,

Natural Killer cells, and cytotoxic T-cells, that are genetically "licensed to kill."

Responsibility for the immune system's surveillance function rests with the cell types termed "antigen-presenting cells" because they can pluck a foreign invader from the bloodstream, prepare it in a certain way, and "present" it to a T-helper cell for evaluation. If the T-helper cell deems the invader "self," that ends the matter; if "nonself," a mighty effort is made to obliterate it, either by the immune system's rapid deployment force of killer cells or—after a delay of one to fourteen days—by a swarm of antibodies.

If it were actually possible to tour a B-lymphocyte, one might see the submicroscopic equivalent of a vast array of factories, storage depots, and railroad yards not unlike that of an automobile assembly plant. Usually on order from a T-helper cell, but sometimes on its own, the B-lymphocyte's factories labor ceaselessly to turn out thousands of copies of its special weapon, the immunoglobulin antibody. Each cell carries the genetic instructions to produce only one type of antibody, but because there is such a huge assortment of B-lymphocytes in our bodies, each producing its own unique product, antibodies come in an almost infinite variety of shapes. The majority of them are the standard models that our "memory" B-cells manufacture and keep in stock for use against such old adversaries as polio virus or tetanus bacillus. On occasion, a foreign organism will enter our blood circulation that the immune system has not anticipated, and in that instance—as will be illustrated shortly—the B-lymphocyte response may be delayed.

Regardless of whether a particular type of antibody derives from a "memory" cell backlog or from a B-lymphocyte specifically activated for that purpose, its manufacture takes considerable coordination. Once a T-helper cell contracts with a B-cell to produce a specific antibody, the work is parceled out to genetic engineers in various nuclear locations. Its variable portion, for example, is usually assigned to genes in chromosome 2 and its joining link to those in chromosome 14. The blueprints for these sections—there are seven in all—are then dispatched to the ribosomal factories in the cell's cytoplasm for assembly. The construction of the weapon's targeting mechanism is especially critical; should it prove defective, the antibody might mistakenly attack a healthy cell instead of a diseased one.

The same standards for precision design and assembly apply to the T-cell's antigen receptor: the other immune system component that comes in a vast assortment of shapes and whose uniqueness is acquired early in a T-helper cell's life. In common with B-lymphocytes, these cells arise from bone-marrow "stem" cells and then travel to the thymus gland, where they undergo rigorous training in how to distinguish "self" from "nonself." Though the process by which this is done is not well understood, mistakes are apparently not tolerated; if a novice T-helper cell erroneously concludes that a particular protein fragment, or antigen, does not belong to the body and sounds the alarm, that unfortunate cell is either destroyed on the spot or imprisoned in the thymus forever. Since the immune system has to be extremely careful about the cells it targets for destruction, less than 1 percent of the immature T-cells entering the thymus ever leave it, and those that do bear a heavy responsibility.

When out on patrol, a T-helper cell works in close cooperation with another member of its flotilla, the macrophage. At forty to eighty micrometers in length, macrophages are three to four times larger than the average lymphocyte and can travel the bloodstream at a maximum speed of 1.7 nautical miles per hour. Macrophages are also phagocytic cells—literally, "cell eaters"—though they serve several other functions as well. When they encounter a suspicious organism or particle, they engulf it, then employ digestive enzymes stored in sacs in their interior to break it down into fragments that fit special molecules essential for T-cell recognition. These proteins are called histocompatibility leukocyte antigens (HLA) by some (because they were originally discovered on white blood cells or leukocytes) and transplantation antigens by others (because their determination and matching, one with the other, is crucial to the success of a human organ transplant). The recent discovery of how these HLA molecules function to screen, prepare, and position protein fragments in such a way that other immune system cells react to them by waving a friendly hello or else going on the attack, represents one of the great triumphs of modern immunology.

Should the protein fragments in the macrophage's belly display an amino acid composition that the HLA molecules "recognize" as suspicious, they will bind to it and migrate to the cell surface. Through chemical signals, such as interleukin-1, the macrophage will then alert all the cells in its flotilla that something might be amiss. In response, a T-helper cell will pull alongside to determine whether the fragment from the foreign invader coupled to the HLA molecules fits its particular antigen receptor.

It may be helpful to point out here that cell membranes are much like key clubs: spherical shells containing many doors, none of which will open to a visitor without the right key in hand. Immune cells are "clubby" in another sense as well; a macrophage, for example, will only deal with a T-helper cell that possesses the same set of HLA antigens as it does. Why such interactions are restricted in this manner is not entirely clear, though it probably has to do with the training in recognizing "self" from "nonself" that the T-cell receives in the thymus. If, in addition to the two types of cells sharing the same HLA molecules, the physical dimensions of the foreign antigen fragment–HLA complex match the T-cell's receptor, it will take the specimen aboard and pass judgment on it. If the verdict is favorable, the alert is canceled; if the specimen is deemed hostile, battle stations sound, chemical signals are dispatched, and the entire flotilla goes into action.

B-lymphocytes display on their surfaces the particular antibody they are genetically programmed to produce. This enables a T-helper cell to "scan" the B-lymphocytes in its vicinity and stimulate the one that makes the antibody it wants. Once activated in this manner, both the T-helper cells and the B-cells undergo what is termed "clonal expansion," reproducing themselves in such large numbers that, within a week or two, they can usually rout the invader.

Our immune defenses represent a formidable force that can readily cope with most external threats to our health. Nonetheless, as in all complex systems, serious mistakes can occur. For example, suppose an influenza virus, a string of genes wrapped in a protein coat, enters a person's bloodstream and somehow goes undetected by his forward patrols. In order to proliferate—which may be all a virus knows how to do—it must first invade an unsuspecting cell and seize control of its reproductive machinery. So it floats around a suitable cell's periphery until it spots an opening, an unguarded dock, so to speak, that its prow can fit into. Then its genes sneak out and head for the cell's nucleus, where, if they're lucky and hit the right spot, they can insert themselves into its DNA coil and subvert it to their own ends. Having missed its first crack at the invaders, the infected individual's immune system must now wait for the virus to proliferate in such abundance that it bursts the walls of the cell it has infested and spills out into the bloodstream. Within minutes, a macrophage will engulf one of the microbes and present it to a T-helper cell, which customarily orders two actions, both fraught with danger: it selects and stimulates a pool of B-lymphocytes to make antibodies against the invader and directs the hunter-killer units in the area to destroy the virus-infested cells.

One of the potential perils of the first course of action is called "molecular mimicry." Concurrent with the population increase that "activated" lymphocytes undergo is a steady rise in the number of genetic mutations their progeny suffer. These are not necessarily bad; some actually make the antibody more efficient in doing the job assigned to it. But others backfire and make it destructive in a different way. An antibody, of course, cannot "see"; it bumps and bounces around until it finds the slot its variable portion was designed to fit. If, however, through mutations in the genes encoding this "claw," it conforms more closely to a molecular structure on a nerve sheath than on a type of influenza virus, a crippling disease, such as multiple sclerosis, can result.

Another source of mistakes in combating a viral invasion involves killer cells. To vanquish the infection, the immune system must not only hunt down the viruses loose in the bloodstream but destroy the cells they've infested as well. A cytotoxic T-cell, for example, is fully capable of destroying a diseased cell by itself, dissolving it away, stem to stern, through use of the corrosive chemicals with which it comes equipped. Yet, compared to viruses or bacteria, diseased cells are much larger and more difficult to spot, and the decision to attack them leaves little room for error, since a misdirected immune system force can severely damage a layer of intestine or the lining of a joint.

In its war against cancer, the immune system faces the opposite problem. Before eradicating a cancer cell, a macrophage, or other cytotoxic cell, must first make tight contact with it by means of another type of HLA molecule that the two cells have in common. If, however, the cancer cell, through mutations that have helped ensure its survival, lacks these adhesion molecules on its surface, the killer cell may fail to get an adequate grip on it. As National Cancer Institute scientists have shown,

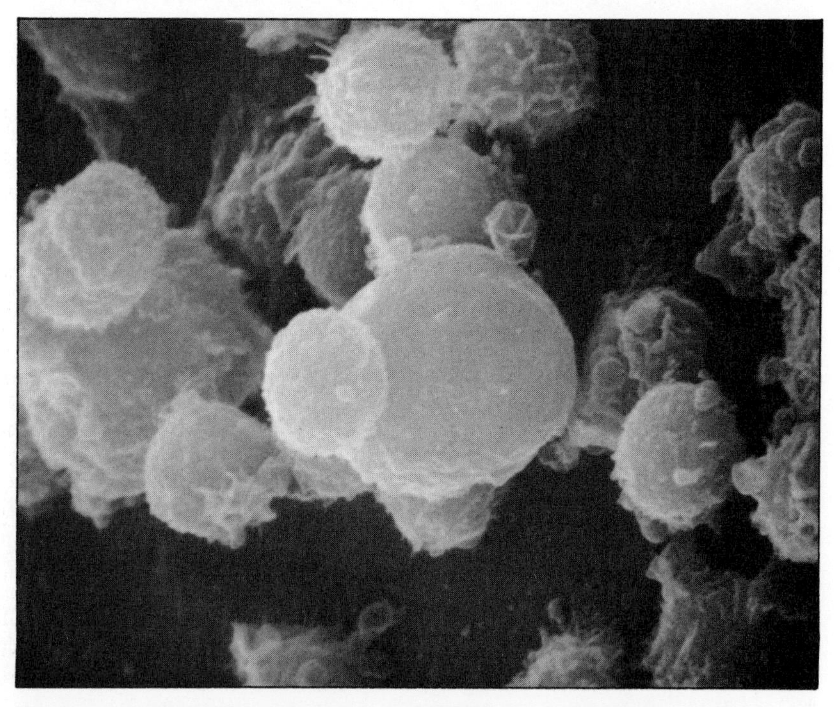

Electron photomicrograph of macrophages surrounding but not attacking a cancer cell

mouse cancers genetically missing these HLA adhesion molecules go unmolested by the animal's immune system until the genes encoding them are introduced into the cancer cells by gene-transfer techniques.

Both Wistar Institute and Swedish researchers have found that one way monoclonal antibodies act against cancer is by recruiting cytotoxic cells, such as macrophages, in the attack. When mixed together before being infused into a patient, thousands of monoclonal antibodies bind to a single macrophage and aid it in finding and sticking to its cancer target. In a series of striking photomicrographs by Zenon Steplewski's team, a flock

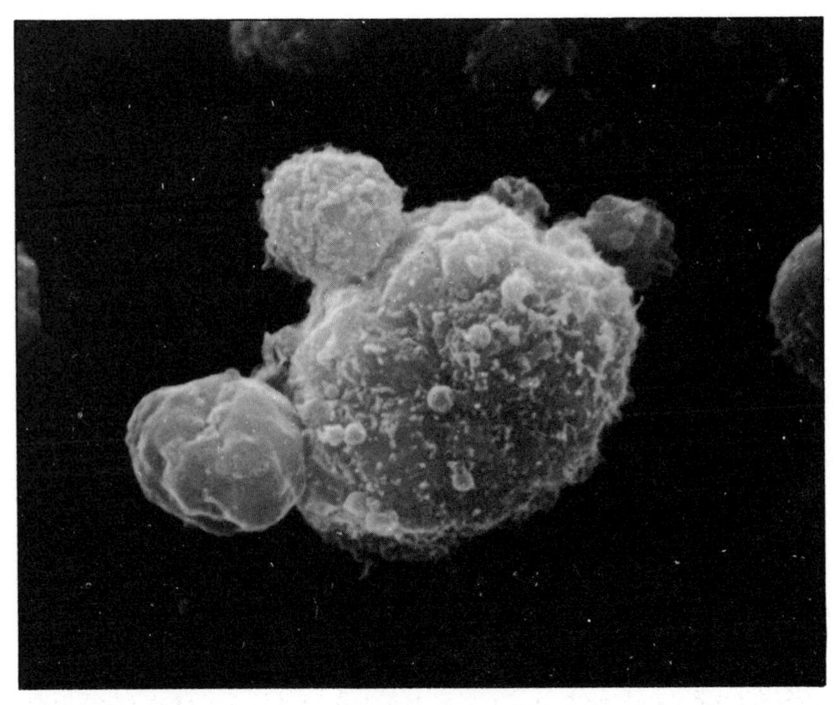

Electron photomicrograph of macrophages primed with monoclonal antibodies attacking a cancer cell. Photos by Professor Zenon Steplewski of The Wistar Institute

of macrophages surround a cancer cell but resist attacking it, possibly because of its slippery surface, until armed by monoclonal antibodies. Moreover, by destroying large numbers of cancer cells and releasing their remnants into the bloodstream, mouse monoclonal antibodies may, at times, "awaken" a sluggish immune system enough so that it starts making its own antibodies against the malignancy.

While around one million Americans are stricken by cancer each year, a far greater number fall victim to autoimmune diseases—literally, an immune system attack on "self." Although exact figures are hard to come by, I would estimate that upwards of twenty-five million Americans currently suffer from such "self-allergic" diseases as rheumatoid arthritis, lupus, juvenile diabetes, ulcerative colitis, multiple sclerosis, and most forms of thyroid disease.

Since we no longer believe, as did many of our ancient forebears, that "to be ill is to be in a state of sin," the mystery remains why some people are afflicted by these ailments and others are not, and how such crippling conditions come about.

To understand the makeup of an autoimmune disease, and many cancers as well, it helps to think of it as a three-tiered structure. The first tier is *hereditary susceptibility*: a latent "time bomb" in the genetic machinery that processes biological information and instructions. Through use of modern gene probes, many of these inherited flaws can now be detected in infants at birth, or even while they are still in the womb, and certain preventive measures taken. The second tier is a *triggering* agent, such as a virus or a chemical, that activates this "bomb." The third, and most crucial, is *immune system malfunction*, brought about, in part, by a lapse in one's physical or emotional defenses.

As shown by Drs. Hilary Koprowski, Michael Oldstone, and others, the mechanism of "molecular mimicry" is a major link between the second and third tiers. Thyroiditis, an inflammation of the thyroid gland affecting 17 percent of all adults, is an example of how these factors combine to produce disease. Suppose a susceptible individual suffers a viral infection that predominantly attacks the glandular tissues of the thyroid. Because of an inherited defect in the immune system cells residing in this gland, they react to the invasion by producing an excess of anti-viral antibodies. And since antibodies are basically killers, this excess turns on and commences to destroy thyroid cells indiscriminately. Or else, through the mechanism of molecular mimicry, an imperfectly constructed antibody attacks both virus and thyroid tissue simultaneously. Either way, the result is inflammation that progresses to complete destruction of the thyroid in approximately one-quarter of its victims.

In addition to providing a provable explanation for what had long been a group of mystifying maladies, these immunological constructs offer hope to the sufferers of autoimmune diseases. If, for instance, it can be clearly established—as Wistar Institute scientists are currently attempting to do—that a specific virus, or small family of viruses, triggers multiple sclerosis in susceptible individuals, then prevention by a vaccine, or amelioration by measures that safely and selectively curtail the immune system's response to the viral invasion, is eminently feasible.

But back to the *why* of it. Regardless of how such diseases as diabetes or cancer come about, why are some stricken by them and others are not? The answer, of course, is that no one knows; no human being yet has been smart enough to figure out the ways of fate, although someday someone might be. In the meantime, our task is clear: to learn as much as possible about the biological clock ticking away in all of us and how to repair it when it breaks down.

7

IN LATE JANUARY 1986, while en route to the international Hybridoma meeting in Baltimore, I joined Jim and Tinka Federline for lunch at a Washington, D.C., hotel. Jim looked better than he felt; over the past few weeks he had developed one of the most dreaded complications of pancreatic cancer, ascites: a progressive accumulation of fluid in the abdomen that causes it to bulge outward like an advanced pregnancy. Long a mystery, the cause of this complication has recently been discovered. Certain cancer cells produce a chemical substance called *vascular permeability factor* that induces the small blood vessels in their vicinity to leak serum through their walls into surrounding tissues or cavities.

Since the onset of his ascites, Jim's abdominal girth had increased a full four inches. Worse yet, the fluid pressing on his stomach restricted its capacity to fill with food, producing intense nausea and a bloated feeling whenever he ate more than small amounts at a time. It also pushed up his diaphragm, reducing his lung capacity and making him increasingly breathless on exertion. Until recently Jim had been jogging several miles every morning to keep fit. He didn't mind so much eating several small meals a day to prevent nausea, but his short-windedness made running impossible and he missed the psychological lift this exercise gave him.

Without saying so, I was even more troubled by this complication; none of the pancreatic cancer patients I had cared for in the past had ever survived more than a month or two after its appearance.

Jim asked what might be done to relieve him of his swelling, and I suggested a trial of combination chemotherapy—knowing he had rejected this option previously but also knowing that the oncologist Jim was seeing in the Washington, D.C., area, Dr. Fred Smith, was an expert chemotherapist. Tinka Federline, having already impressed me on several occasions by how well informed she was on the treatment of gastrointestinal cancers, did so again now by reaching into her purse and taking out a computer printout sheet that, when unfolded, was almost six feet long. Stamped NOT FOR GENERAL RELEASE across the top, it was a highly confidential National Cancer Institute summary of the effectiveness of various chemotherapeutic regimens in the treatment of pancreatic cancer patients nationwide. The results were dismal; a mere 3 percent showed long-lasting improvement.

"Where in the world did you get this?" I asked. Tinka replied that it had come from a high-level National Cancer Institute secretary who happened to have the document available and passed it on. Since that struck me as so unusual, I asked for details.

It was mostly Sally Menigoz's doing, Tinka explained—Sally being Jim's devoted secretary. After numerous phone calls to NCI officials, she finally got to speak to a section chief's secretary who was particularly sympathetic to Jim's plight because her best friend was also battling pancreatic cancer, and slipped

Sally a copy of the summary. With a flick of her eyebrows, Tinka said, "Amazing what you can come up with sometimes, if you dig hard enough."

I could only agree.

We spent the remainder of our time together discussing the potential benefits and real risks of treating Jim with a second course of monoclonal antibodies. Knowing I would be seeing Drs. Koprowski and Steplewski at the Hybridoma congress the following day, I promised to explore the re-treatment option with them.

On the first day of the meeting, my research partner, Dr. Harland Verrill, introduced me to Dr. Jean-Yves Douillard, a French physician and a famous figure in monoclonal antibody treatment circles, whom Harland had met at his Nantes, France, cancer research center. Having begun clinical trials of the Wistar's 17-1-A antibody six years earlier, Douillard had now treated close to two hundred cancer patients with it, more than the total of all other investigators combined. For the last half year he had also been testing a monoclonal "cocktail"—a mixture of four separate Wistar Institute-produced antibodies—on gastrointestinal cancer victims. With Jim Federline in mind, I was anxious to hear his results.

In his early forties, with close-cropped black hair, soft brown eyes, and a bushy beard, Jean-Yves Douillard is a man of gentle manner who speaks fluent English—frequently punctuated by Gallic shrugs. With his round face rimmed by black hair, he rather reminded me of a Franciscan monk. I listened attentively as Douillard reported his results with a four-antibody "cocktail": 17-1-A, 55-2, 77-3, and 19-9. Of the first sixteen patients so treated, two—one with pancreatic cancer and the other with colon cancer involving the liver—had responded with disappearance of all demonstrable tumor masses, and another four patients with shrinkage of their tumors by more than 50 percent (a "partial response").

At the end of the session, I sat down with Zenon Steplewski to discuss Jim Federline's situation at length. Sadly, we agreed that the abdominal fluid Jim had developed was a very bad sign and probably untreatable. Yet, like myself, Zenon had met the Federlines twice now and considered Jim's case "special." If he were to supply me with the three new Wistar anti-cancer antibodies, Zenon mused, would I be willing to administer them to Jim at my hospital?

I hesitated, knowing such re-treatment might not only be dangerous but, unless special permission was obtained beforehand from an FDA official, in violation of federal new-drug regulations. Jim had received his first monoclonal antibody infusion only two months before and so might still have anti-mouse protein antibodies circulating in his bloodstream, which would put him at risk of a severe allergic reaction to a second treatment. Moreover, Dr. Verrill and I had Food and Drug Administration approval to test *only* the Wistar's 17-1-A antibody on our cancer patients. Since the other three lacked what is called an FDA "master file," it seemed extremely unlikely that we would get permission to use them on Jim Federline. And as if these weren't worries enough, Zenon made a further suggestion. Instead of administering the antibodies intravenously, why not infuse a portion of them into Jim's abdominal cavity to attack his cancer directly?

I was intrigued; had it ever been done before?

Zenon smiled uncertainly and shook his head.

To make certain I fully understood the suggestion, I asked whether he proposed infusing antibodies alone into the abdomen or the usual mixture of antibodies plus white blood cells? The latter seemed unlikely, since I knew that billions of white cells pouring into the peritoneal cavity would trigger the same painful inflammatory reaction as that suffered by patients whose appendix had burst or whose stomach ulcer had perforated and would make Jim Federline extremely ill. But, to my amazement, Zenon shrugged and said, "Try it with the white cells. It should work much better that way."

The next morning, over breakfast, I discussed Jim Federline's case with Hilary Koprowski, who also considered Jim "special." While he was not opposed to Zenon's treatment plan, he strongly recommended that Jean-Yves Douillard treat Jim at his Nantes, France, facility, for two reasons: First, French regulations were more permissive than our own when it came to testing new drugs, especially on cancer victims. Second, Douillard, having re-treated dozens of patients with monoclonal antibodies by now, was more proficient in predicting and dealing with allergic reactions. Hilary promised to speak to the French researcher about this, but warned that, because of his busy travel and lecture schedule, Douillard might be unable to accept Jim as a patient any time soon. In that event, would I be willing to re-treat him?

I said I would—and was greatly relieved to learn later that Dr. Douillard had agreed to do it.

WHEN I WAS a medical student in the mid-1950s, my professors predicted that neither the cause nor the cure for most cancers would be discovered in my lifetime. They argued that the secrets of the cancer cell were probably inseparable from those of life itself and so would take another century to uncover.

But now that molecular biologists have cracked the genetic code, solved many of the mysteries of cell-to-cell interactions, and even manipulated DNA molecules to create new life forms, does my professors' prediction still hold true?

Throughout the 1960s and most of the 1970s, it did. Despite the "war on cancer" declared by the United States government in 1971, there were few clinical breakthroughs and the death rate attributable to this disease remained the same. The problem seemed to lie not so much with our troops as in our strategy: before medical scientists could hope to defeat cancer, they had to have a better concept of its nature.

By its most basic definition, cancer is a disorder of cell division. In a metaphorical sense, it is a disorganized and ultimately disastrous attempt on the part of certain cellular genes to return to their freewheeling ways of fetal life. Whether it takes but a single errant gene, a pair acting in concert, or an entire array of genes to trigger this cellular anarchy, is not yet known. But what is clear is that the culprits represent only a very small portion of the genome: the totality of a cell's genetic information. These "oncogenes," as they have come to be called, comprise a family of fifty or so known and probably a few hundred unknown cell-growth regulators that, when expressed or "turned on" inappropriately, cause uncontrolled cell proliferation. It now seems all but certain that they constitute the "final common pathway" of all malignancies.

While the complete oncogene story has yet to be revealed, what has appeared so far makes for an exciting tale, and to tell it properly I must first sketch in a little of what has gone before.

For centuries, physicians surmised that cancer was some sort of infection, like tuberculosis or plague; a slow tumefaction of the flesh produced by some organism entering the body from the outside. There were, of course, problems with that formulation—the main one being that cancers seldom spread from person to person as do other infectious diseases. But the theory persisted for want of a better one.

Then, around the middle of the eighteenth century, the observation of Sir Percival Pott of Kings College Medical School that London chimney sweeps were developing cancers of the scrotum at a rate far greater than what might be attributable to chance led to a radical revision in scientific thinking about cancer. Although nothing was known in those days about the hydrocarbons in soot that can act as carcinogens, Pott correctly deduced that the otherwise rare type of cancer afflicting the young chimney sweeps arose from the repeated rubbing of their thinly protected scrotums against the soot-thick furnace walls. Published in 1775, Pott's finding met with wide acceptance: Cancer wasn't an infectious disease, after all, most medical sages concluded; it was caused by environmental factors. The pendulum of scientific opinion swung so far in this direction that it seemingly got stuck there for well over a century. It wasn't until the early 1900s that a few brave souls—Ellerman and Bang in Europe, Rous in the United States—dared to suggest that the cause of certain rapidly developing cancers in animals might be different.

It is a rare cancer researcher, indeed, who does not know about Peyton Rous. The awarding of a Nobel Prize does much to preserve a scientist's memory, and Rous finally got his in 1966—fifty-five years after he did the actual work that won it for him. The story goes that around 1910 a poultry farmer came to Rous—then a young investigator in the employ of the Rockefeller Institute in New York—with a problem. One of the farmer's prized hens had developed a large tumor and he wanted to know why. To find out, Rous excised the tumor—a soft-tissue malignancy called a sarcoma—implanted portions of it under the wings of a group of hens from the same stock, and lo, they all contracted the same type of tumors. Well, that's interesting, Rous must have thought, but what was doing it? Something in the tumor cells themselves, or in the chemical soup they float in? So Rous took the experiment a step further, inoculating a cell-free solution of the tumor into more birds, and got the same result—leading him to conclude that a transmissible agent (which later came to be called the Rous Sarcoma Virus) was the culprit. So little was known about viruses in those days, however, that Rous's fellow scientists didn't think much of the idea. They argued that even if the growths he had produced in the hens looked like cancers and behaved like cancers, they weren't true cancers at all but merely some sort of weird infection. Rous's contention that they represented viral-induced malignancies met with such widespread disbelief that, to protect his reputation as a serious scientist, he was forced to abandon this research in favor of more acceptable lines of investigation.

With the poor reception given Rous's virus experiments deterring others from conducting similar studies, little was learned about viral-induced cancers until the late 1940s, when the chemical transmitters of genetic information, DNA and RNA, were discovered. Then, with the invention of techniques for studying the molecular makeup of all living organisms, the search got back on track again. In little more than a generation, medical scientists unearthed a string of clues that eventually led them to identify what appears to be the true cancer-causing villains: the oncogenes. One of the earliest and most important of these clues was the discovery of retroviruses—a most peculiar family of viruses whose genetic information consists of RNA rather than DNA. Thanks to a special enzyme, reverse transcriptase, with which Nature had endowed them, they could infect most cells and turn some cancerous. Yet, despite such unusual properties, the retroviruses were long considered mere laboratory curiosities and of no great interest to cancer researchers, even though the Rous Sarcoma Virus was of this type.

Then, in the 1950s, Ludwik Gross of New York's Mount Sinai Medical School, as well as other investigators, found retroviruses that caused tumors in mice and chickens. And at the end of that decade, Scottish veterinarian William Jarrett came across another family member—the Feline Leukemia Virus—which, when injected into cats, not only induced blood-cell cancers in them but caused the disease to spread to other cats living in the same household. With that discovery, interest in retroviruses picked up considerably.

Perhaps, scientists speculated, certain chemical carcinogens from the environment wormed their way into cells already known to be teeming with a wide variety of viruses and activated a latent retrovirus or two. These, in turn, switched on cell-growth-regulating genes in their vicinity in such a way that nothing could shut them off, causing the genes to run wild and a cancer to develop: the so-called Oncogene Hypothesis of Robert Huebner and George Todaro. This came close to what we now know to be the truth, but still wasn't quite right. That became apparent in the late 1970s when it was discovered that the genetic material of most retroviruses, such as the Rous Sarcoma Virus, comprised only four genes—three of them essential to the organism's survival and a fourth, one that the virus had somehow acquired or "captured" from a normal cell, whose sole function, as far as anyone could tell, was to cause cancer. (In 1982, the Lasker Award, the American equivalent of the Nobel Prize, was given to Drs. Hidesaburo Hanafusa, Michael Bishop, Harold Varmus, and Raymond Erickson for their oncogene research. Among the many others who made crucial contributions to this field, Drs. Peter Vogt, Peter Duesberg, Edward Scolnick, and Peter Fischinger deserve special mention.)

Shortly thereafter, cancer researchers got an even bigger shock when they found that this wayward gene wasn't unique to retroviruses at all. By means of an intricate laboratory technique called "molecular hybridization," in which a double-stranded segment of DNA is split and a portion of one made radioactive through isotope linkage and then let loose inside the cell to find its mate, they discovered that the Rous Sarcoma Virus oncogene was actually a copy of a gene present in *all* normal cells. This was puzzling in the extreme: whose gene actually was it—the cell's or the virus's?

Its ownership was finally settled in 1977 when a mutant strain of the Rous Sarcoma Virus lacking the oncogene was introduced into chickens and still induced the same type of tumors. This proved that the gene in question had originally belonged to the cell before being expropriated by the virus. Moreover, it was not just any gene the virus had made off with but a cell-division regulator that, when switched on inappropriately, caused the cell to lose control of its growth.

To simplify a bit, let's imagine that all genes have on-off switches called promoters and repressors, as well as gas pedals and brakes. If anything switches on a gene that should be off—or in the event that it is already on, speeds up the output of the protein it encodes—then cellular chaos can result. It bears repeating that oncogenes are not minor genes, but critically important cell-cycle regulators. (To distinguish them from similar genes hooked to viruses, they are sometimes called proto-oncogenes—"proto" signifying "to go before.") Around fifty of them have been isolated so far, but there are likely to be hundreds more, and the central role they play in the cancer process is beyond doubt.

One may wonder why, if they're such malefactors, the cell has hung on to them throughout eons of evolution. The answer appears to be that the cell simply can't do without them, especially in emergency repair situations when it has to begin to proliferate rapidly again.

That, in brief, is how these cancer culprits came to be identified. Now, researchers have to pin down precisely how the oncogene family operates. Some contend that they aren't the actual villains, but merely accomplices. Cancer may well be a malady of the genes, their argument runs, but what makes them sick in the first place? A chemical? A viral element? A spontaneous mutation in a gene's control sequences? What, in particular, might have triggered the genetic derangements that led to Jim Federline's cancer?

The epidemiology of gastrointestinal cancer is both intriguing and mystifying. From 1950 to 1979, medical statisticians documented a 67 percent decrease in the number of Americans dying of stomach cancer, along with a similar trend worldwide. Although speculations about this steep decline range from diminished exposure to some chemical carcinogen to increased intake of some anti-cancer substance in the diet, its actual cause remains unknown. Over the same period, and for equally obscure reasons, the death rate from cancer of the pancreas has nearly doubled.

A few years ago, a much publicized report in a leading medical journal suggested a link between pancreatic cancer and coffee consumption, decaffeinated coffee in particular. But the results of several subsequent studies failed to confirm this association, and cancer sleuths are still searching for clues to this disease in dim light. Is it a virus, such as the hepatitis B virus that has been strongly incriminated as the culprit in liver cancer? Or a chemical carcinogen, such as one of the nitrosamines that have been linked to clusters of cancer of the esophagus in parts of Asia? Or might it be one or both of these factors in combination with some hereditary flaw?

For the past decade, oncogenes have held center stage in most scientific discussions of cancer causation. By this formulation, no matter what primary event activates them, oncogenes constitute the engine that drives a cell to turn malignant. More recently, however, oncogenes have had to share the spotlight with their opposite numbers: those cell-growth suppressors that for want of a better name are called anti-oncogenes. Though few such genes have as yet been identified, their existence has long been suspected from experiments showing that a cancer cell fused to a normal one often loses its malignant properties. For instance, when an aggressive type of cancer is implanted in a nude mouse that, because it lacks a thymus gland, is immunologically impaired and relatively defenseless against such an invader, the mouse usually succumbs to the malignancy within days. If before being implanted, however, the cancer cells are fused with normal ones, the hybrids so produced neither grow much nor spread—at least initially. Once these hybrids start losing chromosomes—as such cells are prone to do—and, in particular, the ones containing the anti-oncogenes restraining their proliferation, the cancer half exerts dominance over the rest of the nucleus and triggers explosive cell growth.

Genetic studies of a variety of human cancers strongly support the theory that loss of certain cell-growth suppressors—as revealed by missing pieces of chromosomes—are as essential a step as oncogene deregulation in the genesis of many cancers. In retinoblastoma, a highly malignant eye tumor mainly afflicting children, for example, the lack of a portion of chromosome 13 is so consistent a finding that most experts in this disease now believe that the cancer arises, in part, from loss of both copies of an anti-oncogene on this chromosome—one through hereditary error and the other through acquired DNA damage of some kind. Similar genetic analyses of a common type of lung cancer reveal a missing segment on the short arm of chromosome 3 in an exceedingly high percentage of specimens; and in studies just beginning, 20 percent or more of colon cancers show deletions in chromosome 5.

How capable are anti-oncogenes of keeping their opposite numbers from running amok? On the basis of the recent findings of a team of cancer researchers at the Children's Hospital of Los Angeles, the answer appears to be: very capable indeed. Earlier studies had shown that the development of Wilms' tumor, the most common abdominal cancer in children, is associated with loss of a portion of chromosome 11. When the Los Angeles investigators, through the technique of microcell transfer, introduced a normal chromosome 11 into a Wilms' tumor cell line, however, their ability to form tumors in nude mice was completely suppressed.

This demonstration of the potency of anti-oncogenes, along with other experiments to be mentioned, makes clear that cancer is a multi-step process involving imbalances between opposite-acting genes. But what produces the imbalance in the first place? For gastrointestinal cancers this "first hit," as it is called, remains unknown—though the most likely culprits are a virus and a dietary carcinogen, and their favorite targets appear to be the MYC and RAS oncogenes, respectively.

Before getting down to cases, however, let's take a brief look at the four mechanisms through which most oncogenes are believed to operate.

In 1983, the seemingly unrelated research of four groups of scientists suddenly converged to provide a much clearer understanding of the role that the SIS (for Simian Sarcoma Virus) oncogene played in the cancer process. What they found was that the protein SIS produced closely resembled that of the natural cell-dividing stimulus known as platelet-derived growth factor, which supplies the main signal for the trillions of smooth muscle and connective tissue cells in our bodies to divide. This discovery was the first to show that the "extra" or cancer-related gene carried by one of the tumor viruses corresponds to a normal cell-cycle-regulating gene.

Within a year, a second piece of the oncogene puzzle was added when it was found that another such gene, Erb-B, manufactures a malformed growth-factor receptor lacking the keyhole portion of the structure into which the growth-factor "key" is supposed to fit. Further studies showed that this "headless" form of the receptor was actually growth-factor-independent, no longer needing these protein messengers to trigger its signal-generating mechanism and, in effect, acting as a defective ignition switch for cell-dividing machinery that, once turned on, could not be shut off.

A third group of oncogenes, to which the RAS family belongs, is likewise located in the cytoplasm and acts by either spontaneously generating or greatly amplifying the cell-replicating signals that travel along one of the two main trunk lines from cell surface to nucleus. Theoretically, these oncogenes do this by keeping the proteins relaying the signals in an "excited" or actively transmitting state longer than usual.

Finally, there are supervisory genes, such as MYC, that reside in the nucleus, act directly on DNA, and appear to have the final word when it comes to cell division.

How anti-oncogenes work to curb the actions of their rebellious opposites is less clear, though the same four mechanisms, as well as others yet to be identified, are probably involved. To block the effects of excessive growth factor, for instance, they might produce a cell-growth inhibitor, such as beta-interferon, that acts at the membrane level, and perhaps beyond, to restrain the replication of certain types of cells. Inside the nucleus, such genes might maintain order by applying the brakes to a reaction an MYC-like gene had inappropriately induced. Or they might curtail proliferation by directing the cell down the dead-end path of specialization.

With this background, let's examine the RAS family of oncogenes more closely, since newer methods of genetic analysis have found abnormalities in these genes in almost 40 percent of human colon cancers, as well as many other solid cancers and the premalignant growths that give rise to them.

One of the leading authorities on this oncogene, and in particular the mutational events that unleash it, is Dr. E. Premkumar Reddy, who in 1986 joined The Wistar Institute faculty as professor and head of its newly formed section for the study of viruses causing human tumors. During a highly productive six-year stay at the National Cancer Institute, first as a visiting scientist and later as chief of its molecular genetics branch, Dr. Reddy contributed much to our understanding of the molecular structure and function of RAS genes, especially those found in bladder cancers. An unsettled and crucial issue when Reddy began his research in this field was how oncogenes operated. Did they create chaos within the cell by producing too much of an otherwise normal protein? Or did they manufacture normal amounts of a protein whose structure, and therefore function, was so abnormal that it acted at the wrong stage in a cell's growth cycle or for too long a time?

Although they studied several of the known oncogenes, Dr. Reddy and his National Cancer Institute colleague, Dr. Mariano Barbacid, paid special attention to the cellular counterpart of the one carried by the Rat Sarcoma Virus, since they—with others—were finding altered copies of this gene in a significant proportion of animal and human cancers, many of which appeared to be chemically induced.

Earlier studies by three separate research groups had shown that a gene isolated from a human bladder cancer was capable of turning certain mouse cells malignant when introduced into them through the technique of transfection, in which chemically precipitated but still biologically active DNA from one cell is inserted and incorporated into the genetic material of another. The unusually active protein produced by this bladder-cancer gene—later identified as belonging to the RAS family—gave strong indication that it had mutated. To pinpoint exactly where the mutation had occurred and how extensive it was, however, would be a formidable task, since researchers would have to compare, molecule for molecule, the DNA of the cancer-causing gene with that of its normal cellular counterpart, and each was around 6,000 nucleotides long.

Despite its difficulties, Reddy, Barbacid, and their NCI co-workers—simultaneously with the MIT team headed by Robert Weinberg—undertook this investigation in early 1982. What the two groups discovered at about the same time, and published in back-to-back articles in a November 1982 issue of *Nature*, was startling: a single substitution of one DNA molecule for another in a single codon—a change in only one of the three billion or so nucleotides that comprise the human genome—was apparently enough to trigger the transformation of a normal mouse muscle cell into a cancerous one.

Specifically, what the Reddy-Barbacid and Weinberg teams discovered was a "point mutation": guanine replacing thymine in the twelfth codon of the RAS oncogene. This, in turn, leads to a slight change in the order in which the 189 amino acids comprising the protein it produces link up. In place of glycine, valine now occupies the twelfth position in from one end, and as result of this seemingly minuscule alteration, the protein's tumor-inducing properties increase by 100,000 to 1,000,000 times.

Difficult though it may be for some researchers to accept

the concept that a mere 0.0000003 percent change in a gene's composition can result in such disastrous consequences, there exists a precedent. A similar nucleotide switch so alters the hemoglobin molecule that the red blood cells carrying it elongate, clump together under certain adverse circumstances, and cause the deadly disease, sickle-cell anemia.

In a series of experiments with rather frightening implications, Dr. Mariano Barbacid's team has shown that a brief exposure to a chemical known to damage RAS genes can induce breast cancers in a high proportion of experimental animals. The Barbacid group injected a single dose of the chemical nitrosomethylurea (NMU) into the tail veins of young inbred female rats and observed them as they underwent sexual development over the next several months. Since NMU is broken down rapidly in the bloodstream, the animal's total exposure to the intact chemical was no more than a few hours. Yet, following the

For those whose last encounter with the genetic code might have been in high school biology, I offer this brief review. Four different molecules, called nucleotides—adenine (A), thymine (T), cytosine (C), and guanine (G)—in various sequences comprise the seven or so feet of DNA coiled up in each human cell. Since these architectural molecules come in only four forms, yet must provide specific instructions to the ribosomal factories of the cell on how to assemble twenty different amino acids into a protein, Nature deals them out in sets of three—termed triplets or codons—thus making twenty-seven combinations possible. The triplet GTC, for example, codes for the amino acid valine, GCC for glycine, and still others for start and stop signals. But because the symbolic language nuclear genes use in designing proteins differs somewhat from that of the cytoplasmic factories postpubertal rise in their output of estrogen and other breast-stimulating hormones, 90 percent of the NMU-treated rats developed breast cancer. Moreover, genetic analysis of the RAS genes extracted from these tumors demonstrated a nucleotide change in the same position—the critical twelfth codon—as that detected previously in human bladder-cancer cells.

These experimental results, taken together with those of other investigators who have found similar, if not identical, mutations in chemically induced cancers of the skin and lymph glands, clearly indicate that the RAS gene represents the main target for a multitude of chemical and other environmental carcinogens—and this is certainly a major discovery, though it raises a troubling question as well. If a few hours' exposure to a DNA-damaging chemical—a single "hit," so to speak—is all it takes to initiate a cancer in animals, who is to say it doesn't do the same to us? Though this remains to be proven, and the likelihood

assembling them, adjustments have to be made. Accordingly, the instructions encoded in double-stranded DNA undergo certain modifications and editing before being transcribed into single-stranded messenger RNA and dispatched to the ribosomes, where its codons are translated into corresponding amino acids.

This, then, is the genetic code, part one. The second part of the code—whose rules have yet to be completely worked out—governs how the various amino acid chains of the protein fold together to form a functional unit. And as the growing number of physicists who have entered the DNA field in recent years point out, a third part is needed to explain how the three-dimensional protein structures, with their carbohydrate and fat side chains, interact with electromagnetic forces to generate the approximately 120 watts a day it takes to energize a human being.

is that it takes repeated exposure over a long time to get a cancer started, what might one do to reduce this bombardment? Apart from not smoking, what foods, beverages, physical and chemical agents should one avoid? With few exceptions, environmental health experts simply don't know—although it seems possible that, through use of a carcinogen screening system with alterations in RAS genes as the end point, they may soon find out.

EVER SINCE the discovery of viral oncogenes and their normal cellular counterparts, a controversy has raged over whether the runaway action of a single gene is sufficient to turn an otherwise normal cell cancerous. The majority view holds that, at the very least, it takes two separate oncogenes acting in concert to give rise to a cancer, and the results of recent experiments, most notably those of Dr. Robert Weinberg, bear this out. What Weinberg's MIT team found is that the insertion of *both* MYC and RAS genes, though not either alone, into normal rat muscle cells was enough to turn most of them malignant. The reason Weinberg chose these particular oncogenes to test his "two gene" hypothesis is that abnormalities in either the number of copies or activities of one or the other of them have been found in over a third of the human cancers studied so far. Moreover, with MYC acting on a nuclear and RAS on a cytoplasmic level to drive cell proliferation, they frequently appear to be partners in crime.

Unlike RAS, MYC genes, residing as they do behind the nuclear membrane, are better protected from chemical carcinogens and are seldom, if ever, driven wild by mutations. Instead, they're done in by a different mechanism: a mishap in the gene rearrangements certain cellular DNA segments undergo in order to carry out their mission—what one might call a bad shuffle.

By this theory, an MYC gene becomes uncoupled from its on-off control elements through errors in the way breaks in the DNA coil are repaired following these rearrangements and, either on its own or as a result of its influence on other oncogenes, such as the one known as BCL-2, it triggers uncontrolled cell division. Which brings us to a consideration of "jumping genes"—Nature's ad-lib way of adapting to certain environmental hazards—and the problems such genes can sometimes create for a cell.

In 1983, at the age of eighty-one, Dr. Barbara McClintock of the Carnegie Institution's genetics laboratory at Cold Springs Harbor, New York, won the Nobel Prize for her work on "mobile genetic elements"—the formal name for jumping genes. The award, by all accounts, was both richly deserved and long-delayed, since McClintock's discovery of these genetic elements, through experiments involving the changing colors of corn kernels, was made in the 1950s and either ignored or hotly contested for years afterward by diehard adherents to the "dogma of the constancy of the genome." In the traditionalist's view, the instructions of the genetic code were written in indelible ink and could neither be erased, amended, nor altered. McClintock proved them wrong, and as a result we now have a much better matrix in which to fit the emerging pieces of the cancer puzzle.

An example of a human cell whose chromosomes undergo rearranging is the antibody-producing B-lymphocyte. In order to assemble in final form the particular protein antibody it is called upon to make, dozens of its genes have to shift from one chromosomal location to another. Through a process similar to editing a section of film or recording tape, one set of enzymes, the nucleases, snip a gene out of its usual slot, a second set of enzymes transfer it to a new site, and a third set, the recombinases, tie the loose ends together to restore DNA continuity.

Probably not all cells contain jumping genes, and among those that do, such as the B-cells and T-cells of the immune system, the jumping time is limited to a particular phase of their developmental cycle. If by chance—and the chances increase steeply whenever a cell is stimulated to replicate at an accelerated rate—a splicing error occurs that misaligns the control elements of a very active gene, such as one coding for a piece of the antibody molecule, with an MYC-like gene whose activity heretofore had been suppressed, the oncogene's transcriptional engine can get turned on full blast. If, for example, the recombinase enzyme becomes confused by similarities between "signal sequences"—i.e., short strings of DNA molecules it has been specially constructed to recognize and splice—the enzyme may inadvertently bind genetic elements from two different chromosomes together: an unnatural and unpredictable situation. The more frequently such mistakes occur, the more likely that the "gear shift" promoter or "gas pedal" enhancer of a highly active gene will end up next to an MYC gene and take control of it. Or an oncogene might simply become detached from its repressor: the lock on its promoter and enhancer sequences that keeps it in check. Either way, the oncogene swings into action, cloning multiple copies of itself, boosting its rate of transcription into RNA, overproducing the protein it encodes, and making a general mess of the cell's growth cycle.

Though other ways exist to activate MYC in human cells,

most investigators would agree that chromosomal derangements represent the main mechanism—and that Carlo Croce's Wistar Institute team is largely responsible for their discovery.

I first became aware of Carlo Croce's work in 1980, when his photograph appeared on the cover of *Medical World News* above the caption "Monoclonal Antibodies Go Human." Ever since Köhler and Milstein in England created the first hybridoma, researchers had raced to see who could reduplicate the feat with human-cell equivalents and, in a photo finish with Stanford University researchers, Croce won. By successfully hybridizing a human lymphocyte immunized against measles virus with a human cancer cell, he harvested a batch of monoclonal antibodies specific for a portion of the virus. Perhaps even more important are the discoveries of Carlo Croce's team in the fast-moving field of chromosomal cartography: the assignment of one, or a small cluster, of the 100,000 or so genes of the human genome to one of the twenty-three pairs of chromosomes that comprise its major divisions.

In his early forties, Carlo Croce is medium tall and trim with wavy black hair worn long in the back, broad cheekbones, and highly animated features. Though a meticulous investigator, Croce appears to possess the verve and instincts of the successful gambler. He gambled when, having graduated with highest honors from the University of Rome Medical School, he chose research over the practice of medicine. He gambled again when he entered the then undeveloped and potentially arid field of molecular genetics. And he spends much of his leisure time gambling for high stakes in art objects: as a self-taught expert in the works of sixteenth- and seventeenth-century Italian artists, he buys, sells, or trades their paintings and sketches.

Reminiscing about his early days at The Wistar Institute,

Croce says, "I spoke very little English, didn't know my way around or how to get things done, and except for Hilary Koprowski—who, after all, was my boss and not someone I could run to every time I needed something explained to me—I was completely on my own. So I sat in my cubbyhole of an office and read a lot until Hilary—who, long before Milstein and Köhler created their first hybridoma, foresaw a lot of useful information coming from cell-fusion experiments—got me interested in them. So for the next five or six years, I fused one type of cell with another to see what the hybrids could, or couldn't, do after they started losing chromosomes. Then, after a year's sabbatical at the Carnegie Institution in Baltimore to learn the latest techniques, I became a full-time gene mapper."

In 1979, Croce succeeded in assigning the locus, or location, of the gene cluster encoding the heavy-chain portion of the human antibody molecule to an end segment of chromosome 14. In addition to its two heavy chains, an antibody consists of a pair of light chains of one of two distinct types, called kappa and lambda; within two years their genes, too, were located. Working independently, both Croce's team and the one headed by Philip Leder of Harvard Medical School mapped the kappa-type light-chain genes to chromosome 22. A third team of British and American researchers then mapped the lambda gene to chromosome 2. The importance of this work became obvious in 1982, when Croce, in collaboration with National Cancer Institute scientists, discovered that, though the normal location of the MYC gene was on chromosome 8, in certain cancer cells it had moved to chromosome 14. With this crucial clue in hand, a whole array of previously puzzling findings involving odd shapes and exchanges between chromosomes, as well as oncogenes, suddenly began to make sense.

To start with, there were the genetic mysteries surrounding Burkitt's lymphoma, a disease that usually begins as a jaw tumor and spreads rapidly throughout the body. Named after the British surgeon who first described the condition in 1950, Burkitt's lymphoma mainly afflicts young African children and is a peculiar malady indeed. Although clearly a cancer of wild-looking white blood cells, it exhibits several features of an infectious process as well, especially in its concentration in certain geographic locales and in its association with a specific virus. Laboratory studies reveal that over 90 percent of its victims show evidence of recent Epstein-Barr virus infection—surprisingly the same virus that causes the usually benign illness, infectious mononucleosis.

Strange as the half-infectious, half-cancerous disease described by Burkitt behaves clinically, its appearance under the microscope is even more unusual. Through special staining procedures that can delineate forty to fifty distinct regions of each chromosome, it has been discovered that virtually all these cells have undergone a chromosomal rearrangement of some kind, most commonly an exchange of tiny segments between chromosomes 8 and 14. It was only after Croce's team mapped the MYC gene to chromosome 8 and the antibody heavy-chain genes to 14, however, that the significance of this change became clear: by moving from one location to the other, MYC had escaped from the DNA sequences keeping tight rein over its expression and into the embrace of the more active, or at least permissive, regulators of the heavy-chain genes. Moreover in the 15 to 20 percent of cases where, instead of landing on chromosome 14, MYC ends up on 22 or 2—the location of the kappa and lambda light-chain genes, respectively—the same set of circumstances holds true.

Subsequent studies of hundreds of cases of Burkitt's lymphoma confirmed that translocation of the MYC-containing segment of chromosome 8 to one of the three aforementioned sites occurred uniformly and that, though this change of itself might be insufficient to trigger a malignancy, it sets the stage for one. A strong immunological challenge, such as malaria or a severe viral infection—anything that forces these genetically unstable cells to replicate rapidly—might then cause one of them to explode into a cancer clone. And even though Burkitt's lymphoma is extremely rare outside East Africa, the disease, by being the first proven instance of the cause-and-effect relationship between MYC gene deregulation and human cancer, has yielded up a secret of great theoretical importance.

To account for the weak spots in chromosome 14 that might have prompted the heavy-chain gene's control sequences to inadvertently turn MYC on, one has to go back a bit into B-cell history.

Since a newborn's immunological defenses take several weeks to develop, the infant at first is totally dependent on the antibodies derived from the mother's circulation. Then, once his immune system realizes what it is up against in the outside world, it begins to manufacture its own assortment of antibodies. Because of the myriad foreign invaders the B-cells have to contend with, they must shuffle their thousand or so antibody-making genes around considerably to meet every contingency. For example, to fashion the appropriate heavy chain for a particular antibody, four separately designed components—the variable, diversity, joining, and constant segments—have to be assembled. Approximately 100 genes code for the variable, 50 for the diversity, 6 for the joining, and 9 for the constant portions of the heavy chain, making $100 \times 50 \times 6 \times 9$ or 270,000

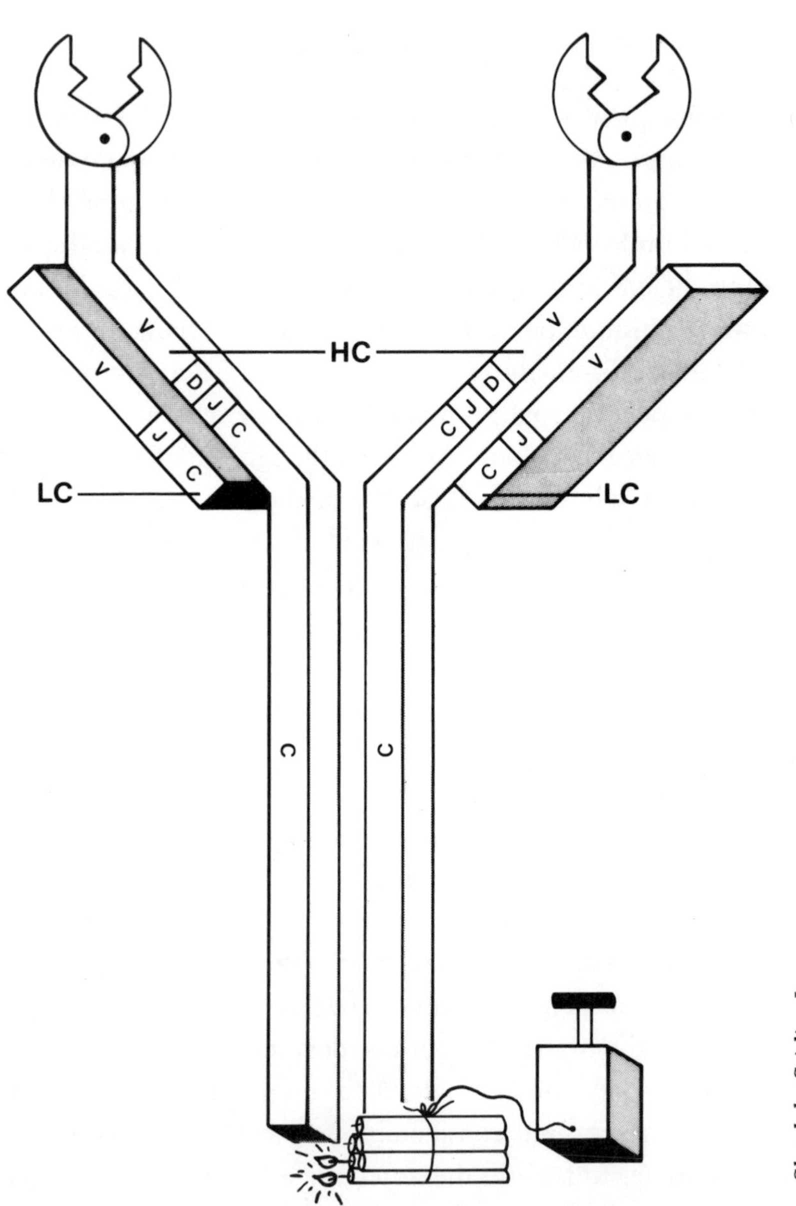

Sketch by Stirline Lacy

THE ANTIBODY MOLECULE

HC	heavy chain	D	diversity region
LC	light chain	J	joining region
V	variable region	C	constant region

different combinations possible. Multiply this by the thousands of possible variations in the manufacture of the light chains and one has a repertoire of a billion or more different antibodies.

Though the vast majority of these genetic rearrangements occur early in the life of the B-cell, still more changes must take place as the cell matures and prepares to release its antibodies into the bloodstream. Through a process called "isotype switching" that modifies its heavy chain, an antibody is further equipped to carry out its given mission, whether this be the removal of allergy-producing molecules from the skin or the killing of cancer cells.

Thus, in its contests with outsiders, the immune system has the cards to win almost every hand. Yet through bad shuffles or misplays of the sort that places an oncogene next to an antibody-producing gene, a terrible irony can occur: as if an evil spell had been cast over it, a B-cell gives birth to its archenemy, the cancer cell.

Chromosomal translocations consistent from cell to cell have now been detected in all the various types of B-cell lymphomas and in 70 percent of its leukemias—the majority of them involving MYC or the oncogenes known as BCL 1 and 2 (for B-cell lymphoma/leukemia)—and are considered the crucial step in the genesis of these malignancies.

Much the same sort of misalignments occur in T-cell cancers as well. Although this type of white blood cell does not make antibodies, it, too, must undergo extensive gene rearrangements in order to fashion the surface receptors necessary to take up and analyze all the various protein fragments presented to it by macrophages. As the inevitable consequence of this constant reshuffling, its DNA develops weak spots, especially in the region on chromosome 18 that contains the genes' coding for one chain of the T-cell receptor. Thus, in most T-cell tumors studied by Croce's group and others, MYC genes have been found in the midst of receptor genes instead of where they belong on chromosome 8.

How much guilt does MYC bear for the occurrence of B-cell, T-cell, and possibly many other types of human cancers? To my knowledge, no consensus currently exists on this issue, and since scientific "truth" is defined by whatever a preponderance of the experts agree on at the time, no definitive answer is possible. Even so, I cannot help wondering how MYC would fare if it were actually possible to put the gene on a witness stand and subject it to a prosecutor's cross-examination.

Although the scientific jury is still out on MYC, Carlo Croce predicts that it will acquit the nuclear gene on at least two charges: archvillainy and premeditation. While it appears true that MYC's deregulation plays a crucial role in the development of virtually all blood-cell tumors and that a low-grade malignancy cannot progress to a metastasizing cancer without its help, it is also true that all normal cells probably have to go through MYC in order to divide.

If I've gone on a bit long about jumping genes and malfunctioning enzymes, it is because of my belief in their fundamental importance. For just as the discovery of mutations in RAS genes provides researchers with a major link between chemical carcinogens and certain cancers, this new knowledge does the same for viruses and cancer.

In 1979, the National Cancer Institute's Dr. Robert Gallo proved conclusively that a specific virus causes a specific human cancer. Originally calling it the HTLV virus after the disease it induces, human T-cell leukemia, Gallo subsequently added the number 1 to its designation upon discovering that it was closely related to a family of viruses that causes other T-cell cancers, as well as AIDS. What remains to be explained about this group of human tumor viruses, however, is how—since they carry no known oncogenes—they transform cells into cancers.

As previously touched on, Carlo Croce favors the view that these viruses, as well as others yet to be identified, exert their pernicious influence primarily by producing a population explosion among certain immune system elements: B-cells in Burkitt's lymphoma and T-cells in acute T-cell leukemia. This, in turn, increases the likelihood that a DNA splicing enzyme in one of these rapidly proliferating cells will make the sort of error that activates an oncogene—giving that cell a selective growth advantage over its neighbors. And that's more or less where things stand until another growth stimulus, such as an infection, causes the cell's enzyme system to make an even worse mistake, especially one that unleashes MYC. In support of Croce's theory, experiments have shown that the insertion of MYC oncogenes into B-cells infected with the Epstein-Barr virus results in their rapidly turning malignant.

A second mechanism whereby viruses can deregulate oncogenes directly, instead of by chromosomal rearrangements, has recently come to light and involves those DNA sequences that regulate gene action. To review a bit: most cell-cycle genes are turned on by protein messengers in the form of growth factors or hormones that act on its promoter and enhancer regions. Some of these genes then produce proteins whose sole mission is to move down the chain of command to activate other genes. Applying this schema to viruses, researchers have shown that, soon after infecting a cell, a virus activates a set of "early stage" genes to make proteins that migrate to the control sequences of an array of "late stage" genes and, through a poorly understood process called *transactivation*, turns them on. Should, however, these transactivating proteins get misdirected or lost and, instead of attaching to a viral gene, end up on a cell-cycle-regulating gene, major problems can result. Worse yet, some of these viral proteins appear to act similarly to those produced by certain of the oncogenes. For example, one of the proteins elaborated by the HTLV-1 virus stimulates a sharp increase in an infected T-cell's ability to manufacture growth-factor receptors. In contrast to normal cells of this type, such a cell is much more likely to respond to whatever growth-factor molecules it receives excessively, if not uncontrollably, and so serve as the forerunner to a leukemia.

So much then for the blood-cell cancers that, though they afflict 70,000–80,000 Americans each year, often fatally, are far less prevalent than solid cancers of the lung, breast, bowel, and skin. For purely technical reasons, the systematic study of the chromosomal changes such tumor cells undergo as they progress from low- to high-grade malignancies has lagged behind that of the lymphomas and leukemias. But those that have been done show clearly that genetic abnormalities of all types abound, especially chromosomal deletions that may contain anti-oncogenes.

For instance, in a third of liver cancers induced by hepatitis B virus infection in animals susceptible to this human virus, chromosomal rearrangements involving MYC genes have been observed. This finding has prompted Carlo Croce to speculate that the same sort of DNA splicing errors discovered in Burkitt's lymphoma may occur in liver cells infected with the hepatitis B virus and give rise to cancer.

Which at last brings me around to a question I raised earlier but did not adequately address: what caused Jim Federline's cancer? Although the answer lies beyond the frontier of what is currently known about pancreatic cancer, a plausible scenario might be the following: Either a carcinogen-induced mutation in an RAS gene or a viral infection that, through the mechanism of transactivation, turned on some other oncogene delivered the "first hit" to one of Jim's pancreas cells, altering it in such a way that it became "partially transformed" or premalignant. Upon subsequent replication, most of the progeny of this unstable cell died of genetic malformations or else were destroyed by the immune system. A few, however, survived, only to suffer a second hit: another mutational event or a chromosomal rearrangement that activated an MYC-like nuclear oncogene. Though still in a premalignant phase, such cells have now been "immortalized"—meaning that their "biological clock" has been so deranged that they can reproduce themselves endlessly but cannot develop any useful function; they are doomed to perpetual immaturity. Nonetheless, such local factors as lack of an adequate blood supply to furnish food and oxygen keep their rate of growth within bounds until a third and decisive event—most likely, a chromosomal deletion—leads to further oncogene activity and unrestrained proliferation.

The fate of the normal cells abutting and in the near vicinity of this cancer clone now rests in the hands of the immune system's surveillance and killer cells, whose efficiency, in part, depends on the coordinated central nervous system discharges and brain chemicals that help to sustain their effort.

But before leaving the realm of scientific fact to speculate on whatever sub rosa arrangements may exist between the immune system and the brain, there are a few points worth making about the practical, as well as theoretical, value of the chromosomal studies that Wistar Institute scientists have helped to pioneer.

Both in composition and in clinical behavior, the various types of lymphomas and leukemias, of which there are dozens, differ widely. Some run so indolent a course that treatment can safely be delayed for years; others ravage and kill a victim within weeks of its apparent onset. Thus, each case, and its management, must be assessed individually. For this purpose, monoclonal antibodies and gene probes have been developed to enable physicians to precisely identify the type of cell that has turned malignant, predict its course, and select the optimum therapy for it. The Wistar Institute team headed by Carlo Croce and Kay Huebner has mapped the chromosomal locations of an even dozen oncogenes so far and, through study of the DNA breakpoints that commonly and consistently occur in a variety of malignancies, hopes to discover and fashion gene probes for dozens more.

Future efforts along these lines may well encompass the precise diagnosis and classification not only of established disease but of *potential* disease as well. How? Through the study of both blood cells and blood products—the former to spot characteristic gene amplifications, chromosomal rearrangements, or missing anti-oncogenes; the latter to detect elevated serum levels of oncogene-related proteins. For those individuals shown to be at risk, preventive measures to destroy premalignant clones, or else halt their progression to full-blown cancers, might then be instituted. Though the form such preventatives will take remains uncertain, they might include anti-cancer vaccines of the type known as "anti-id"—the prototype of which is currently being developed by Dr. Dorothee Herlyn—or highly selective monoclonal antibody infusions.

10

AT THE OUTSET of their journey, Tinka Federline's main worry was how well Jim would tolerate the eight-hour flight from Washington to Paris, since he became increasingly uncomfortable sitting for long periods. But Jim quickly eased her anxiety on that score by falling asleep shortly after takeoff and sleeping soundly the rest of the way. Then there was the more serious concern raised by Zenon Steplewski during their last phone conversation.

Zenon reported that Jim's recent blood sample remained positive for anti-mouse antibodies, which was good in a way—it showed how active his immune system was—but bad in another, since it made re-treatment more perilous. Steps to minimize the chances of an allergic reaction would be taken in the form of skin tests: injections of very diluted solutions of each of the antibodies to be used under the skin of his forearms and the sites observed for redness or swelling; should Jim react positively to any of the antibodies, the treatment would have to be canceled. But such precautions were not foolproof, Zenon warned; Jim could still suffer a life-threatening allergic reaction when the same molecules were introduced directly into his bloodstream.

In a subsequent phone conversation between Jim and Douillard, the French physician repeated the same warning. Yet Jim remained determined to go through with the treatment; to make it an all-or-nothing gamble. So it was on to Paris and, after a change of airplanes, two hundred more miles southwest to Nantes, an ancient Gallic city of around 300,000 inhabitants at the mouth of the Loire estuary.

When the Federlines asked about the weather there, Dr. Douillard had told them it was chilly but that it hardly ever snowed. Yet as their jet began its descent to the Nantes airport, all they could see out the window were snow-topped roofs and ice-glazed roads. For the first time in years, a heavy snow, followed by freezing rain, had fallen the day before to practically paralyze the region. Tinka refused to view the storm as any sort of omen, but it did make for a bleak introduction to the Loire-Atlantique department's capital city and Dr. Douillard's hometown.

Fortunately, there was a lone taxicab outside the terminal building and Jim and Tinka hurried to it. From a week spent in Paris two summers before, they had picked up a smattering of French, though not nearly enough to carry on a conversation with the elderly cabdriver, who knew no English whatsoever. The locals spoke a little Spanish, since Spain was a neighbor, a little German if absolutely necessary, but no English. Tinka made their destination known by printing the name and address of the University Hospital on a piece of paper and handing it to the driver, who grunted an acknowledgment and got the car underway. Grumpy though he appeared, the cabdriver was not unkind; he not only drove the Federlines to the hospital but guided them to the reception desk, which, without any signs indicating its location, was on the second floor. A woman in white uniform greeted them cordially but, having received no advance notice of Jim's arrival, had no idea what they wanted. Worse yet, the receptionist acted as if she had never heard of Dr. Douillard. This seemed incredible at the time, but later became understandable when an English-speaking nurse explained that physicians of Douillard's stature were routinely addressed as *Monsieur le Professeur*, not *Docteur*, a lesser title, which had confused the receptionist until Tinka printed his name on a piece of paper. "Ah! Professeur Douillard! . . . Doo-yard," she exclaimed, trilling its French pronunciation. After consulting with her superior, she escorted the Federlines to a room in the research wing of the hospital, where Douillard would join them shortly.

The University of Nantes's medical school and hospital occupy twin nine-story structures linked by an enclosed second-story ramp. From the outside, the modern buildings look much the same as their U.S. counterparts; what differences exist are primarily on the inside and particularly in the no-frills patient rooms. The one to which Jim Federline was assigned was typical: tiny and bare-walled, it contained a rigid metal bed, a small table with two high-backed chairs, and a sink; a toilet was several strides down the corridor. Putting down their suitcases, Jim and Tinka sat on the bed and stared searchingly at one another. With its broken tiles and peeling paint, the room looked so dingy that they couldn't help wondering if this was the sort of place where Jim ought to undergo so risky a treatment.

Dr. Douillard, whom they had spoken to on the telephone but never met, arrived within the hour and quickly put the Federlines at ease with his friendly manner, excellent command of English, and obvious competence. He examined Jim, explained in painstaking detail the procedure they would follow the next morning, and then personally drove Tinka and Jim to Nantes's best hotel.

Reassured by Douillard's presence and his vast experience with monoclonal antibody therapy, the Federlines checked into the Hôtel de France, were assigned a comfortable room, ate dinner in its restaurant, and lingered at their table until nine p.m., when Jim returned to the hospital and the room in the research wing where he was to spend the night.

At six a.m., a silent orderly brought him a breakfast tray of standard French hospital fare: a bowl of warm milk, a hard-crusted roll, and a pat of butter. Unable to break the roll with his hands, Jim left the room in search of a knife and spotted a hospital employee in the kitchen dunking his roll in the warm milk to soften it. So that was the trick! he thought, and hurried back to his room to devour his breakfast.

Tinka joined him there around 7:30. Soon afterward Dr. Douillard appeared with a short, bespectacled French nurse named Marci, her long hair braided and fastened at the top of her head. The veteran nurse spoke English, and when she was not busy with other duties, would serve as their interpreter. If ever an American couple needed a few French friends, Tinka told me later, this was the time, and Marci turned out to be an endearing one.

An hour later, two young French nurses wheeled a washing machine-sized, IBM-manufactured blood-cell separator into Jim's room. They thoughtfully brought along two pocket dictionaries, handing the English-French edition to Tinka and keeping the French-English one for themselves. This is how we'll communicate, they indicated with gestures, and when Marci was not around, it was how they did.

The nurses prepared Jim for the leukophersis procedure by

inserting a long plastic catheter into a forearm vein and another into the large femoral vein in his right groin and connected both to the machine. The blood-cell separation process took three hours; at its conclusion, six billion or so of Jim's white cells, tinged by some red cells, had been isolated in a plastic bag and transported to Douillard's laboratory to be incubated for an hour with the four-antibody "cocktail" he would soon receive.

Before Douillard returned with the mixture, a cart of emergency drugs and equipment was wheeled into the room, its implications clear: even though the antibody skin tests applied to Jim's arms earlier had proved negative, there could still be dangerous reactions from the "cocktail," and if so, drastic countermeasures would have to be taken.

Douillard surprised Jim by informing him that, instead of infusing half the antibody–white blood cell mixture intravenously and the other half into his abdominal cavity, as originally intended, he had decided to put it all into his abdomen, where it would likely do the most good. Jim did not bother to ask Douillard if this would increase the pain he expected to suffer afterward; he just assumed so and told him to go ahead.

Douillard numbed a half-dollar-sized spot on Jim's lower abdomen with a local anesthetic, made a small skin incision with a scalpel, and inserted a large-bore plastic tube into the fluid-filled abdominal cavity.

There followed a moment of extreme tension as the intravenous tubing connecting the plastic bag to the catheter in Jim's abdomen reddened along its length with the antibody–blood cell mixture. Everyone present realized that the chances of an adverse reaction—pain, hives, chills, vascular collapse—were considerable at this point, yet Jim remained the calmest person in the room.

For the next hour, as the antibodies in the bag drained into his abdominal cavity, the Federlines and Douillard chatted casually about a wide range of subjects, with no complaints coming from Jim. By midafternoon he had six billion white blood cells roaming his peritoneal cavity, but he still felt no discomfort. Even after Douillard warned again that antibody-coated lymphocytes were killers and it was only a matter of time before they began killing cancer cells and he started hurting, Jim could not help feeling jubilant. The danger of an acute allergic reaction had passed and he was confident he could handle the pain. In a small way, he had made medical history: no cancer patient with a bellyful of fluid had ever been treated in this way before; should the treatment work well on him, others with the same hellish disease might benefit.

Over the next several hours, Jim felt nothing worse than an occasional abdominal twinge and, with Dr. Douillard's permission, left the hospital for the hotel and another excellent French dinner.

He woke the next morning with a sore abdomen, particularly at the site of the catheter placement, but otherwise he felt well. When Tinka joined him at the hospital, he was pleased to learn that their new friend, Marci, had met her at the hotel, driven her to Nantes's famous Cathedral of St. Pierre to pray, and kept her company the rest of the evening.

Before releasing Jim from the hospital, Dr. Douillard gave him an injection of a powerful painkiller to keep him comfortable on the journey back to the United States. They arrived home the afternoon of February 26—Jim's forty-second birthday. Together with their two daughters, they celebrated it joyfully.

Excruciating stomach pain woke Jim around dawn the next morning. His entire abdomen felt ablaze. He swallowed a couple of painkillers but they helped not at all. By midmorning, he was in such agony that Tinka had to give him a shot of Demerol. When, two hours later, he asked for another shot, she pleaded with him to go to the hospital, but Jim wouldn't hear of it. He refused even to see his doctor. If the killing of cancer cells was what was causing his pain, then the more, the better; it seemed a fair price to pay and he would simply have to endure it.

But the pain not only persisted for the next week, it grew so intense one night that Tinka was on the verge of calling an ambulance. Had he developed the intra-abdominal abscess Douillard had warned about? Was he dying? Jim did not deny this was the worst torment he had ever experienced but he insisted on waiting until morning—after the kids went off to school—before going to the hospital. Reluctantly Tinka gave him his third Demerol injection of the night and after a while he drifted off to sleep.

When Jim woke the next morning, he was amazed to find the pain completely gone; his belly didn't even feel sore. He remained pain-free the entire day, and the next morning, putting on his pants, he noticed they were roomier in the waist. Using his tape measure, he discovered that his abdominal girth had shrunk almost two inches from a week before. He felt so good he was tempted to go for a jog, but Tinka restrained him. Wait a couple of days to get back your strength, she urged, and Jim did, though by the end of the week he was jogging two to three miles every morning.

His oncologist, Dr. Fred Smith, was pleased, if not astounded, by Jim's obvious improvement. To document it, Smith obtained a CAT scan that showed the amount of fluid in Jim's abdomen had decreased by two-thirds; without a doubt the unprecedented treatment he had received in France was working.

Jim phoned me to tell me the good news, and throughout the rest of the day I thought about him and the courageous struggle he was waging against an all but overwhelming disease.

Another battle had been won in what still looked to be a long, tough war.

11

IN APRIL 1986, urged on in part by Centocor, the Malvern, Pennsylvania, biotechnology corporation manufacturing The Wistar Institute's 17-1-A antibody, Hilary Koprowski convened a two-day meeting of the leaders of the eight groups treating gastrointestinal cancer patients with this immunological weapon.

Almost from the inception of monoclonal antibody technology, researchers assumed that these molecular missiles, like polyclonal antibodies before them, could kill cancer cells in the test tube or in nude mice bearing human tumor implants. But experience had taught that, essential as these experiments are, they are often poor predictors of the way the human body, a far more complex organism, will respond. In Hilary Koprowski's view, the immune system is all too often "a loser from the very beginning" in a cancer war; either the cancer goes undetected by the victim's immunological defenses, suppresses them in some unknown way, or overwhelms them. So the question he hoped to have answered was whether monoclonal antibody therapy could bolster an immune system on the verge of defeat enough to turn it into a winner?

In perhaps the most often quoted of Claude Bernard's observations, the famous French physiologist once wrote, "A scientist does not do experiments to confirm his hypotheses, but to control them!" And Hilary Koprowski has never lacked for bold new hypotheses. Some he could put to the test himself; others he could assign to members of his staff; still others died of neglect. Yet Hilary could not rid himself of the notion that monoclonal antibodies might well represent the new approach to the treatment of cancer that everyone in the field agreed was needed. The outcome of the pilot study, conducted from 1979 to 1982 by Henry Sears, Zenon Steplewski, and Dorothee Herlyn, in which three of twenty cancer patients responded dramatically to 17-1-A antibody infusions, fueled his belief in this form of therapy, and at a Hybridoma meeting in Los Angeles in 1983, he recommended the procedure, hoping to stimulate some in his audience to try it. Instead, a verbal shoot-out erupted between Koprowski and officials of the Food and Drug Administration over whether mouse-derived monoclonal antibodies, in any dose, form, or combination, were ready for human use at all.

Hilary left the meeting determined to prove the skeptics wrong. Despite the risk to his reputation should further trials of the 17-1-A antibody prove harmful, or a total flop, he encouraged eight cancer treatment centers worldwide (including my own in Flint, Michigan) to join forces with The Wistar Institute to test the antibody's effectiveness in hundreds of patients over the next several years.

The National Cancer Institute conducts such "therapeutic trials" all the time, and when they fail, as many understandably do, little is said. But for an independent institute, lacking its own treatment facility, to undertake so extensive a study represents a steep gamble for all involved, with the possible exception of the patients themselves, who often have the least to lose. Cancer of the pancreas is an almost certain killer: 25,500 Americans fell victim to it in 1986, and 24,000 died. Cancer of the stomach afflicts an equal number of people annually and kills a high proportion of them. The most common of the gastrointestinal malignancies, colon cancer, strikes around 140,000 Americans each year and, in a sense, their fate is as unpredictable as a coin toss: approximately half are cured by surgery, and despite all available countermeasures, half suffer fatal recurrences within five years.

When informed of these frightening statistics, and their bleak prognosis, the vast majority of gastrointestinal cancer victims feel they have little to lose by submitting to experimental therapy, such as monoclonal antibodies, especially since the initial treatment is virtually free of side effects. Put another way, we who treat these patients know that by so doing we are not depriving them of more effective remedies. Yet in making this therapy available, we invariably create a dilemma for ourselves. Whereas the patients we would like to treat are those with what is termed a "low tumor burden," the ones we are implored to treat often have such advanced disease and depleted immune defenses that nothing is likely to cure them. So, while our hopes for each patient run high, our expectations are modest: to buy time until something with greater curative power comes long.

Before leaving for Philadelphia and the 17-1-A conference, I phoned Jim Federline to find out how he was faring. In his usual upbeat manner, Jim said okay, though on further questioning I learned that he had had to cut short a family vacation on his sailboat because of persistent stomach pain and that his abdominal fluid was reaccumulating. I promised to discuss his case with Drs. Koprowski, Steplewski, and the other monoclonal antibody experts attending the Wistar symposium to see what else might be done, and to get back to him.

Our phone conversation unsettled me. Sooner or later, we all reach the end of the line, I thought gloomily, and Jim seemed close to that terminus. In the records I kept, Jim's condition was classified as "stable." Reluctantly I changed this to "rapidly progressive," and fought off despair only by reminding myself of the many crises Jim had already surmounted. More "miracles" might still be possible, and if so, the 17-1-A meeting was the place to look for them.

Among the hundred or so researchers attending the Wistar symposium, "The Immunodiagnosis and Immunotherapy of Gastrointestinal Cancer," were the leaders of the eight treatment centers: the University of Nantes, the University of Munich, and Stockholm's Karolinska Institute in Europe; the Fox Chase Cancer Center, the National Cancer Institute, the Hurley Medical Center, the University of Alabama Hospital, and the University of Nebraska Hospital in this country.

In his welcoming remarks to the gathering, conference chairman Zenon Steplewski pointed out the rapid progress made since the first 17-1-A-producing hybridoma was created at The Wistar Institute a mere seven years earlier. Meenhard Herlyn provided further perspective by outlining the eighteen laborious steps—from the immunization of mice with the colon-cancer antigen to the harvesting, purification, and testing of each batch of antibodies for safety and potency in laboratory animals—that had to be performed before such antibodies were suitable for human use. Left unmentioned was the approximately $7,000 it cost the Institute to make enough of them for the treatment of a single patient, since none was expected to pay for it. This was experimental research, after all, and until quite recently patients were never asked—at least officially—to bear the expense.

Several reports were then given on the structure of the 17-1-A antibody, how it and other monoclonals lent themselves to the early diagnosis of gastrointestinal cancer through blood tests or isotope scanning, and the ways in which antibodies might be made better cancer killers by linking them to radioactive isotopes, cancer chemotherapy agents, or poisons, such as the incredibly deadly castor bean extract, ricin, a single molecule of which can kill an entire cell, or diphtheria toxin.

The second day of the symposium was devoted to reports of patient responses by the eight treatment groups. A total of 376 people had received 17-1-A infusions—271 for therapy of their cancers and the remainder for phase-one toxicity testing. Of the "treated" group, approximately a third had benefited to some degree. Twenty-six patients had achieved a "cure" (absence of demonstrable tumor) or else a disease-free interval lasting over two years. Sixty-nine others were sufficiently stabilized to survive at least an additional year—a result that, without a "control," or no-treatment, group (nearly impossible to obtain, since most cancer patients demand and deserve treatment) with which to compare it, is uninterpretable. Nevertheless, it was a fairly consistent finding from group to group. For example, of the first thirty patients, all with advanced cancers, my Hurley Medical Center team treated with the 17-1-A antibody, seventeen were alive a year later and two were free of disease—less than a spectacular success, to be sure, but at least equal, if not superior, to that achieved by any other form of therapy.

It also bears repeating that, in contrast to the complications that often ensue from chemotherapy, few 17-1-A recipients have suffered significant side effects from this treatment, and none at all from the initial infusion. Thus, for as long as this therapy proves effective, it does not diminish the "quality of life." This safety factor applies only to the first treatment, however; as discussed previously in Jim Federline's case, the more often the patient is re-treated with the 17-1-A antibody, the greater his chances of suffering an allergic reaction to the mouse protein. And although such reactions have occurred far less frequently than first anticipated, and none have been fatal, they limit further monoclonal antibody treatment.

Sidestepping the serious, though not insurmountable, allergy problem for the moment, what can one make of 17-1-A therapy? Clearly, it is not a cure for most gastrointestinal cancers, but might it be the beginnings of one? *Spontaneous* cures do occur in cancer patients, though they are extremely rare, their estimated incidence being one in one hundred thousand cases. Nonetheless, the question remains: Are the 17-1-A findings statistically significant? Even more problematical, does the statistician exist who would dare analyze them in the absence of "controls"?

Except for the fact that all the test subjects had gastrointestinal cancer and received at least one 17-1-A infusion, the extent of their disease, as well as the totality of their treatment, differed considerably. Some underwent the procedure of leukopheresis, in which five to ten billion of their white blood cells were collected and incubated with the monoclonal antibody prior to infusion, and others did not. Their ages, general states of health, the sizes and aggressiveness of their tumors, also varied widely—as did the unquantifiable, though perhaps crucial, factor: the will to live. To complicate matters further, many had undergone radiation, chemotherapy, or other immunological treatments, such as gamma-interferon, before their 17-1-A therapy. In short, they were a very mixed patient population—which makes for poor science and poor peer acceptance of results. Yet this is frequently the case in clinical cancer research, and where the scientific and humanistic concerns of the researcher collide.

The pure scientist strives for an experimental model with as few variables as possible, which is often achievable in glassware or in laboratory animals. But what is the physician to do when entrusted with the care of a patient suffering an all but incurable disease, such as pancreatic cancer? How, if experimental agents are available to him, can he reconcile his obligations to do everything possible for his patient and to advance scientific knowledge at the same time? Some take the "shotgun" approach, reasoning that it makes more sense from the start to try everything that's fairly safe and the least bit promising and, should the patient improve, figure out what worked best later on. The clinical research directors of drug companies decry this method as wasteful of time and resources and, in the main, they are right, though it does make for sound sleep at night.

The more academic physician-researcher, though sympathetic to the shotgun approach, pursues the scientific goals of accuracy, objectivity, and reproducibility more assiduously. Like the veteran combat commander, he must be willing to risk lives and sustain "acceptable losses" in the hope of victory. Hence, he prefers the randomized, double-blinded experimental design, though its only reward is often a number, a P value, that determines the degree of statistical significance. Such phase-three studies may sound heartless, and in some instances they are, but whether performed early or late in a new treatment's development, they are a necessity; without them, the Food and Drug Administration, with rare exceptions, will not sanction an agent for general use, and should it prove truly lifesaving—as I believe 17-1-A therapy *with modifications* will be—fewer Americans will benefit.

Of the two great unknowns the 17-1-A group had set out three years earlier to explore—was the treatment safe? was it effective?—the answer to the first was yes, and to the second: in some patients, especially those with early disease. But as with virtually all scientific inquiries, the answers merely generated more questions and, by general agreement, the most pressing of these involved dosage. Medical history is replete with instances of effective treatments that failed early trials and were nearly abandoned because they were administered in too low a dose. The L-dopa treatment of Parkinson's disease is an example. Once it was discovered that victims of this devastating malady lacked the nerve impulse-transmitting chemical dopamine in certain of their brain centers, neurologists hoped that, by feeding its parent substance, dopa, to such patients, the dopamine contents of their brains would be replenished and their suffering relieved. But it was only after the daily dose of L-dopa was increased from a few hundred milligrams to several grams that this drug proved an effective therapy.

Perhaps an even more telling example—and one that, until the dosage issue is settled, will continue to haunt me—occurred in the early days of penicillin when the antibiotic was first tested in patients with bacterial endocarditis, a life-threatening disease in which bacteria entering the bloodstream attack an already damaged heart valve and, if unchecked, cause multiple organ failures and death. Because of its spectacular success against pneumonia, physicians hoped penicillin would cure bacterial endocarditis, and eventually it did, but not until they raised the dose from thousands to many millions of units per day.

Due to uncertainty over how well humans would tolerate mouse monoclonal antibodies, as well as their limited supply, a 17-1-A dose range of 200 to 500 milligrams was arbitrarily selected, with the vast majority of patients receiving around 400 milligrams. Was this amount sufficient to destroy a solid cancer that, in many instances, consisted of five hundred billion or more cells? Hilary Koprowski had his doubts; he conjectured that 17-1-A therapy might have failed in many patients because their cancers were exposed to the antibody for too short a time and that larger doses at more frequent intervals might produce better results. Should it turn out—as evidence currently being gathered by Dr. Hakan Mellstedt of Sweden's Karolinska Institute suggests it will—that, on a weight-for-weight basis, it takes as much antibody to cure a patient of gastrointestinal cancer as it does to eradicate a similar type of tumor implanted in a mouse, then a total dose of 8 to 12 grams would be needed. In fact, Mellstedt's finding in a colon cancer patient that, even after he had received 4.8 grams of 17-1-A over a twenty-four-day period, surgical biopsy showed that less than 50 percent of his tumor was saturated with antibody, has prompted the Swedish investigator and others to administer even larger doses to their patients, such as 500 milligrams every other day for 48 days.

But even when the proper amount of an antibiotic is given for a bacterial infection, it may still fail to cure because the bacteria rapidly become resistant to it. They do this through genetic rearrangements that produce "resistance factors"—a phenomenon whose recognition led to the discovery of DNA and RNA in the late 1940s and recombinant DNA technology in the 1970s.

By a similar defense mechanism—the manufacture of proteins that extrude potential poisons from the cell as fast as they gain entry—cancers become impervious to many of the chemotherapy drugs, and agents are currently being developed to block the actions of these transport proteins.

Moreover, clever though bacteria may be in the art of survival, the cancer cell, whose nucleus provides it with thousands of more genes, has a far greater repertoire of tricks, so please bear with me while I attempt to explain the vexing problem of "tumor heterogeneity": the differences in the cells comprising a cancer. Although most malignancies are believed to arise from a clone of identical cells, the pressures of survival force so many mutations that, by the time the cancers are clinicaly detectable, they usually constitute a hodgepodge of different cell types, each with its own distinctive set of surface proteins known as tumor-associated antigens.

On first meeting a patient I plan to treat with monoclonal antibodies, I try to explain the principle behind the treatment by asking him to picture the surface of a cancer cell as a forest with a grove of trees within it that set it apart from its neighbors. A colon-cancer cell, for instance, may sprout a type of oak that normal cells do not. Why? Because the oncogenes giving rise to the cancer produce such an abundance of its particular protein that some appears on the cell surface, where it either acts or merely accumulates. It is precisely this "botanical" difference that enables monoclonal antibodies to tell one cell type from another and target only those displaying "oak trees" for destruction. Accordingly, a surgical or biopsy specimen from a patient's tumor is almost always tested against the 17-1-A antibody, or one closely akin to it, before treatment is offered. Sound simple? When looking down the barrel of a microscope at cancer cells on a glass slide, it is. But within the body these same cells can conceal, change, or shed some of their surface structures and so elude an antibody attack.

To try to eliminate this "hard core," multiple monoclonal antibodies, each directed against a different target or "tree," have been administered together—as in the four-antibody "cocktail" Jim Federline received in France—and are sometimes successful. Or the patient might be pretreated with alpha-IA or gamma-interferon, since these substances have been shown to increase the number and variety of distinctive antigens on a cancer cell's surface.

But even when the antibodies hit every one of their targets, there is still another factor to consider: their killing mechanism. The usual way antibodies destroy viruses or bacteria is to swarm all over them and blow tiny holes in their membranes until they literally drown in the salty water surrounding them. Antibodies accomplish this in two stages: first, their variable portion—their "claw"—locks onto the organism; then their constant portion, the demolition end of the molecule, attracts a series of proteins, collectively called complement, from the blood circulation to trigger its explosive charge.

But, as mentioned previously, the destruction of a cancer cell, millions of times bigger than an antibody, is a far more formidable task and usually requires a cytotoxic mechanism: the killing of one cell by another. What happens here is that the variable, or combining site, of the antibody binds to the cancer cell and its constant end to a killer cell, such as a macrophage or cytotoxic T-cell. Then the killer cell releases corrosive chemicals to dissolve the cancer cell.

A new strategy that is beginning to be tested is to increase the number of killer cells in the circulation by giving the patient interleukin-2 along with monoclonal antibodies. In an *autocrine*, or self-stimulating, manner, interleukin-2, a chemical released by T-cells, hooks around and reattaches itself to many of its parent cells, inducing them and other types of killer cells to proliferate. It thus plays a key and perhaps commanding role in immune system responses. Recombinant DNA technology has recently made vast quantities of this immunological weapon available to researchers and they have tested it in hundreds of cancer patients—around a third of whom responded favorably, provided they could tolerate its severe side effects. When administered to patients in high enough amounts to be effective, interleukin-2 causes a mysterious "capillary leak syndrome" in which fluid seeps from these tiny blood vessels into surrounding tissues, such as the lungs. However, by treating patients with much smaller doses of interleukin-2 in conjunction with monoclonal antibodies, it may be possible to get the same, or better, results with fewer complications.

Another, even riskier, strategy is to try for a "first-strike capability"—an antibody attack that would wipe out 90 percent or more of the cancer cell population in a single blow—by linking a powerful radioactive isotope to the monoclonal so that, instead of needing to deliver a high concentration of antibody to the cancer cell in order to achieve a kill, merely one or two would do. Moreover, armed with a "superbomb" isotope, the antibody missile would demolish not only its target but the several layers of cancer cells surrounding it. Monoclonal antibodies linked to such isotopes as iodine 131 or bismuth 212 have, in fact, been tried with varying degrees of success in the treatment of lymphoma and leukemia patients.

What looks to be a safer and even more promising approach is to attach antibodies to the plant poison ricin, as Berkeley,

California, researchers have done—the underlying principle being that the ricin molecule consists of two chains of amino acids, alpha and beta; the beta chain affords the molecule entry into a cell and the alpha chain kills it by paralyzing certain of its vital functions. What the Berkeley group therefore did was to split the ricin molecule and replace the beta chain, which enables the alpha chain to enter *any* cell, with a monoclonal antibody, giving it access to only certain cells.

It is a clever concept that, in preliminary trials, has proven relatively safe and effective—with about a third of a series of fifty-five patients suffering from the deadly skin cancer, melanoma, showing tumor shrinkage following a *single* ricin-conjugated antibody infusion. Although few patients have so far been "cured" of their disease by this method, it appears to be only in its infancy. Newer laboratory techniques to produce selective mutations in the large ricin molecule have led to second- and third-generation immunotoxin-antibody constructs that promise to be far safer and more effective for human use.

Which brings me to what until now has been a major drawback to monoclonal antibody therapy. Regardless of how much, or in what form, these antibodies are given, the likelihood is that, to achieve maximum benefit from them, most patients will have to be re-treated, possibly several times—putting them at increasing risk of an allergic reaction. To minimize this danger, molecular biologists are creating a new breed of antibody, half mouse, half human, that some call "chimeric"—a chimera being a mythological creature with a lion's head, a goat's body, and a serpent's tail—and others "designer" antibodies, since they are, in fact, "custom-made." What has prompted this bit of laboratory legerdemain is the finding that, of the two main functional units of an antibody—its variable, target-seeking portion and its constant, demolition portion—it is the latter that, when of nonhuman origin, can provoke an allergic reaction. Hence, a near-ideal antibody to use therapeutically would be one with its variable region from a mouse and its constant region from a human—and such hybrids have actually been assembled by The Wistar Institute's Peter Curtis, as well as others. How? Through an intricate technique known as transfection, in which the human genes encoding the constant portion and the mouse genes the variable portion are inserted into the nucleus of an antibody-producing cell. Though untested in human subjects as yet, in laboratory animals these "designer" antibodies have performed as well as their mouse-mouse counterparts.

Thus, for every strategy the cancer cell employs to foil the immune system, there appears to be counter-strategy, which is why the prospects for immunotherapy look so bright. Nonetheless, as was apparent to all who attended the 17-1-A symposium, such therapeutic weapons as "chimeric" and ricin-conjugated antibodies are well into the future; what might they do now for the thousands of patients stricken by gastrointestinal cancer every year?

A separate study conducted and reported on by Jean-Yves Douillard provides one direction. Dr. Douillard and his French colleagues tested the ability of a single 17-1-A infusion to prevent recurrent colon cancer in a series of patients who had just undergone "curative" surgery but whose cancers, under the microscope, were found to have penetrated the width of the bowel wall to the fat or lymph nodes in the immediate vicinity, though not beyond that point. By current statistics 50–60 percent of such patients will suffer recurrences of their cancers within five years. Yet, in Douillard's series of ninety patients—half of whom received 17-1-A shortly after their surgery and half did not—only

14 percent of antibody-treated patients developed recurrent disease over a mean follow-up period of two years, compared to 40 percent of the untreated "controls." Should these preliminary findings hold up over the full five-year follow-up period of the study *and* be independently confirmed, the lives of 30,000–40,000 colon-cancer victims a year could theoretically be saved in this country alone.

Both during Douillard's talk and Hilary Koprowski's summation, Jim Federline's case, and his dramatic, if only temporary, response to the intra-abdominal infusion of monoclonal antibodies, was mentioned. At the end of the meeting, I asked Zenon Steplewski, "Can anything more be done to help him?" And Zenon, who except for Polish political matters tends toward optimism, shrugged and said, "Something, yes. We *must* do something, but just what I don't yet know."

Shortly thereafter, Zenon obtained a liter of Jim's abdominal fluid from his Washington, D.C., oncologist, isolated the cancer cells floating in it, and tested them against every monoclonal antibody available to him.

In late May, Zenon phoned to ask if I would be willing to re-treat Jim with a new monoclonal antibody cocktail, despite the danger that he might suffer an allergic reaction. My research partner, Harland Verrill, happened to drop by my office at this time and I repeated Zenon's request to him. Harland's face went through the same sort of contortions that I'm sure mine had moments before; then, sighing deeply, he nodded.

I told Zenon we would.

12

ON JUNE 4, 1986, Jim and Tinka Federline returned to Flint. When my secretary informed me of their arrival at my office, I felt a sense of unease, if not foreboding. We had treated twenty-eight patients with monoclonal antibodies by now—some for the second time—without a single complication. But Jim would be the first of our patient group to receive the antibodies for the third time, and I had to wonder whether our luck would hold. From our phone conversations, I'd learned Jim had lost ten more pounds and was now receiving supplemental nourishment through an indwelling plastic tube called a Portacath inserted into a vein under his left collarbone and emptying into his vena cava. I also knew he was severely anemic. So, as I went to greet him in the waiting room, I expected to see a wan, emaciated man.

Actually, Jim did not look very ill. As always, he was neatly dressed and groomed and his smile came so easily that, for the moment, my anxiety over his re-treatment lessened.

They had enjoyed a pleasant flight, Tinka reported. Then she said something that touched me deeply: Flint had come to represent a place of hope for them.

I took Jim to my examining room. He looked pale and frail, though not wasted. Nor did he show much in the way of abdominal fluid, since his Washington, D.C., oncologist, Fred Smith, had drawn most of it out with a needle a few days before. Underneath the fluid cushion, however, I felt a rock-hard mass filling much of the upper abdomen that I knew to be his cancer.

As I helped Jim up into a sitting position, I felt obliged to warn him that, should any of the allergy-determining skin tests to be done on him in the morning prove positive, I'd have to cancel the treatment. "And even if they're negative," I went on, "you still run the risk of a severe reaction."

"So?" Jim replied. "What have I got to lose?" Though his shrug was accompanied by a smile, the glint of determination in his eyes was diamond-hard. I remembered reading a medical report by a British allergist that claimed hypnosis could abolish positive skin reactions in patients previously shown to be allergic to the test substances. Could Jim's force of will do the same? At that moment, I tended to think it could.

But later that afternoon, as Harland Verrill and I stood outside Jim's hospital room discussing our plans for the morrow, I grew less confident about the way things might go and decided to have an anesthesiologist present at the start of the infusion. Should Jim suffer the sort of severe allergic reaction that sent his vocal cords into spasm, he would need immediate intubation—the insertion of a plastic tube into his windpipe, followed by respirator-assisted breathing—and an anesthesiologist would be far more adept at this procedure than I. Harland agreed wholeheartedly with the precaution, and so did Tinka when I mentioned it to her later—though she felt compelled to tell me that, months earlier, Jim had signed a "living will," a legal document that forbade his doctors from trying to keep him alive by artificial means should he fall terminally ill and be unable to think or decide for himself. Jim had done this primarily to spare her from ever having to make the agonizing decision to turn off life-support machines, Tinka explained, and at my nod of agreement, added cheerfully, "Don't worry. Nothing bad's going to happen tomorrow. Jim'll do just fine!"

Much as I hoped Tinka's prediction would prove accurate, there came a moment the next morning when I had to hold tightly to my professional calm. As the first few cubic centimeters of the monoclonal antibody–white blood cell mixture entered Jim's bloodstream, I saw him stiffen to suppress a shiver. Fearful that he was having a chill, the first sign of an allergic reaction, I seized the stopcock on the intravenous tubing and was about to terminate the infusion when the pleading look in Jim's eyes stopped me and I put my stethoscope to his chest instead. Though he was breathing rapidly, almost panting, the movement of air through his lungs sounded sibilant and unobstructed. I listened next to his pounding heartbeat—which gradually slowed, along with my own.

The rest of the infusion mixture went in as smoothly as if all it contained was a harmless sugar solution. The following morning Jim and Tinka flew back home to prepare for a month-long family vacation at their beachhouse and for whatever might come next.

Later that day, I admitted to the hospital a patient we had treated with monoclonal antibodies four months before and, because the cancer deposits in his liver were spreading, planned to treat again the next morning.

Unlike Jim Federline, this middle-aged gentleman was sad-eyed and slow-moving, worn down by constant pain, and even though he expressed gratitude for our refusal to give up on him, I had the distinct impression he was going along with the re-treatment more for the sake of his wife and daughters, who hovered near him, than for himself. With his cancer-riddled abdominal CAT scan in mind, I thought that he would probably die soon, and from the resigned, if private, looks he gave me I gathered he thought so, too. It was scarcely the first time a cancer victim had conveyed such a feeling to me. It was their way of being realistic and, at the same time, relieving me of the burden of pretense or false cheer. Yet it never failed to fill me with a sense of inadequacy. Someday, I thought, we of the medical profession will be able to treat advanced cancer patients as successfully as we now treat those with hypertension or peptic ulcers. But until a cure for the common cancers becomes as great a national priority as putting an American on the moon was a generation ago and containment of the AIDS epidemic is now, that day did not figure to be any time soon.

As for the gentleman in my office, I felt we understood each other. If nothing else, the treatment he would get in the morning, whether it benefited him to some slight degree or not, would at the least help to sustain the hopes and prayers of his family.

That evening, as I watched a news telecast and yet another debate between scientists over the wisdom of a "Star Wars" type of defense and the militarization of outer space, I got to thinking about a different sort of war in a different sort of space—the one raging inside Jim Federline's abdomen that only the Washington, D.C., surgeon who had operated on him the year before had ever witnessed firsthand. This led me to rise and take down from my bookshelves John Milton's epic poem *Paradise Lost*, whose study and subsequent appreciation was forced upon me in college, and whose vivid imagery comes to mind whenever I ponder, without conclusion, the morality of men killing other men in war.

The scene I remember best is in Book XI: Adam and Eve, having been duped by Satan into eating the forbidden fruit, are about to be banished from the Garden of Eden and God sends the angel Michael to prepare Adam for what the future holds for his and Eve's descendants. Michael transports Adam to a broad plain where large armies, some on horseback, some on foot, are slaughtering one another in ferocious combat. Having never before seen death, Adam is horrified. There follows a passage meant to convey Milton's message that such is the price men must pay for the glorification of war. It might apply equally well to those genetic derangements leading to oncogene activation and the induction of many cancers.

In tears at the relentless slaughter before him, Adam turns to the angel Michael and asks:

> "O *what are these*
> *Death's ministers, not men, who thus deal death*
> *Inhumanely to men and multiply ten thousandfold?*"

Michael replies:

> "*These are the products*
> *Of those ill-mated marriages*
> *Where good with bad were matched, who of themselves*
> *Abhor to join, and by imprudence mixed*
> *Produce prodigious births* . . ."

Rereading this scene with its striking application to "oncogene marriages" reinforced my belief that, just as electron orbits exist within planetary orbits, much of what goes on in the microworld resembles the macroworld, and evil forces exist at all levels of life. With my proclivity to anthropomorphize, it also led me to wonder how, if it were possible to communicate with a cancer cell, it would justify its existence. It might, I would guess, claim that, despite the hardships it must endure—the perennial shortages of food and oxygen, the immune system patrols intent on its destruction—it is a superior type of cell whose only wish is to reseed the host with its progeny: cells that need never die.

Reproductive immortality is, in fact, a constant feature of cancer cells. Unlike their normal counterparts that, under the influence of tissue organizers, mature, specialize in some way, and ultimately die, cancer cells can go on replicating endlessly—though they must pay a terrible penalty for it. Lacking the cell-cycle maturation and differentiation factors that enable normal cells to develop and do useful work, the cancer cell remains frozen in infancy forever.

Nonetheless, the question arises: Do cancer cells hold the secret for a longer life span for us humans? Improbable though this might seem, there is a precedent: it was, after all, from another leading killer, disease-producing bacteria, that scientists first learned of the existence of DNA and, more recently, of restriction endonucleases: those marvelous enzymes that split DNA strands at predictable junctions and so make recombinant DNA technology possible. And in a recent address to the staff of the Worcester Foundation for Experimental Biology in Shrewsbury, Massachusetts, the eminent physician and essayist Dr. Lewis Thomas had this to say about the role viruses may have played in speeding our evolution: "Maybe to get all the way from the earliest bacteria to the human brain in so short a time as 3.5 billion years, you need more ways than sex allows for the exchange of genetic ideas. Maybe that's what viruses do for a living."

Is it possible that, in its own hazardous and obscure way, cancer can teach us something important, too?

To consider just what, let's take a Zen Buddhism approach, one of its tenets being that each force in the universe has a counterforce—the positive with the negative, the yin with the yang—and, ideally, they should be in perfect balance with one another. Thus, positrons pair with electrons, matter with anti-matter, birth with death.

What, then, is cancer's opposite?

Dr. Vincent Cristofalo, Wistar Institute professor and gerontology expert, thinks that it might be senescence. For years now, he has accumulated evidence that strongly suggests cancer is the flip side of aging.

13

SOMETIME IN the coming years, in one way or another, medical scientists will perform the first successful gene transplant to correct a mistake of Nature, a missing or malfunctioning bit of genetic information that has incapacitated its victim. When that happens, it will represent as great a triumph for molecular biologists as landing a man safely on the surface of the moon did for aerospace engineers. It will be the grand culmination of a research effort that began in the 1940s with the discovery of DNA, progressed in the 1950s with the elucidation of DNA's twin-stranded structure, and accelerated in the 1970s with the development of monoclonal antibody and recombinant DNA technology.

With the availability of gene transplants to treat many of the victims of hereditary disorders, such as a sickle-cell anemia or cystic fibrosis, the list of *totally* incurable diseases will shrink considerably. But even if medical science succeeded in eliminating all causes of disease, few of us would live beyond the age of one hundred and ten. As Shakespeare wrote, "The fault . . . is not in our stars, but in ourselves . . ." Our genes are simply not programmed for it.

Efforts to understand and retard the aging process are as ancient as the medicinal use of roots and herbs and date back at least five thousand years to the Mesopotamian seers who promulgated the myth that a thorny plant with the powers of rejuvenation grew at the bottom of the sea. Although the search for such a magical substance has now shifted from the ocean floor to the laboratory, it endures with vigor and with rising expectations for success. For just as molecular biologists are creating new life forms almost daily through advanced technology, the possibility exists that they will soon discover ways to make human beings more durable.

Gerontology, the study of the aging process, is not only a relatively new branch of science but one that until recently has been so rife with quackery that it has had to struggle mightily to overcome a bad reputation.

That pitchmen for the secrets of the ages, and the ageless, still abound and continue to leech off the work of serious investigators is clearly indicated by the steady procession of books written and promoted by unblushing authors who profess new knowledge but seldom validate anything more profound than P. T. Barnum's adage: "There's a sucker born every minute." Yet studies of the life cycle of cells at the molecular level are progressing so rapidly that, within another decade or two, medical science may well find a way to extend our life span beyond its current maximum of about one hundred and ten years.

The Wistar Institute—due largely to the accomplishments of two of its former faculty members, Drs. Paul Moorhead and Leonard Hayflick, and the team now headed by Dr. Vincent Cristofalo—ranks among the world leaders in the gerontology field. Tall, husky, and amiable, Vincent Cristofalo is a soft-spoken man in his early fifties who has helped raise six daughters and in his spare time keeps fit by playing pickup basketball on a court near the Institute. A biochemist by training, he joined the Wistar staff in 1963, drawn there by David Kritchevsky's energy metabolism studies and the fascinating findings of Moorhead and Hayflick's cell-culture experiments. As Cristofalo relates the story, the Wistar Institute's entry into the gerontology field occurred almost by accident.

One of Hilary Koprowski's main objectives on assuming the directorship in 1957 was to determine the feasibility of growing viruses in human cells, rather than in monkey cells. His reasons were twofold: since certain viruses, such as those of the polio family, showed a clear preference for human over animal flesh, the yield would likely be greater, and the vaccines derived from these cultures less allergenic. Hilary assigned the project to two newly recruited staff members, Moorhead and Hayflick, who quickly realized its difficulties. Like orchids, human cells are very finicky about the soil, climate, and sundry other conditions in which they'll grow. The scientific literature was unhelpful; few attempts to grow human cells in glassware had hitherto been made and fewer still had succeeded—and so they proceeded by trial and error, finally isolating a human lung-cell line, dubbed WI-38, that grew well in culture dishes. But try as they might to keep the strain alive on serial subcultivation—the procedure in which some of the cells are transferred to new dishes to give them more room to grow—they couldn't do so for indefinite periods. After about fifty population doublings, the vast majority of the cells stopped proliferating, degenerated, and died.

The Wistar scientists were stumped. Though consistent from experiment to experiment, these results ran contrary to the prevailing view among cell biologists that individual cells in culture were immortal; only hierarchies of cells that, together with nerves and blood vessels, constitute a tissue or organ ever aged and died. This misconception was based on the outcome of a *single* experiment, but one of incredibly long duration, carried out by no less a personage than Nobel laureate and Rockefeller University professor Dr. Alexis Carrel. An innovative biologist and cell culturist, as well as a famous surgeon, Carrel began growing normal, nonmalignant chick heart cells in a test tube in 1911 and kept them growing for the next thirty-four years, at which time he voluntarily terminated the experiment. Since no chicken had ever been reported to live beyond thirty years, Carrel concluded that, if properly maintained, cells grown outside the body reproduce endlessly, and fellow scientists, none of whom was particularly eager to repeat such a lengthy experiment, accepted this as fact. To this day, no one knows for certain how Carrel got these results, though the usual explanation is that each time he nourished the cultures with chick embryo extract, he unwittingly introduced viable new cells into their midst. Whatever the circumstances, no scientist since has been able to get similar cells to undergo more than thirty-five population doublings or survive longer than forty-four months.

Knowing none of this at the time they were struggling to keep their WI-38 cells alive, Moorhead and Hayflick were about to admit defeat when it dawned on them that what they were actually witnessing was normal cellular senescence. In another of the great scientific discoveries that began as heresies, they went on to prove that cells were mortal, after all; that they, like us, must eventually grow up, specialize, and find gainful employment, though the price such toil exacts is death.

Subsequent studies of cells from sources as diverse as fruit flies and turtles have shown that the number of times they can renew themselves in culture dishes is directly proportional to the life span of their species. For example, embryonal cells from a mouse, whose life span is 3.5 years, double around twenty times under similar conditions, and those from tortoises that live a century and a half or more, ninety to one hundred and twenty-five times. Moreover, when Hayflick took WI-38 cells that had undergone eight population doublings, froze them in liquid nitrogen, and reconstituted a portion of them monthly over a thirteen-year period, they all went on to replicate a total of around fifty times, strongly suggesting that some intrinsic mechanism—a biological clock, so to speak—keeps track of these molecular events.

Thus, the answer to the question of how cells know how old they are appears to be that they "count" replications. Exactly how, no one knows, but perhaps a timer of some sort, a clock wheel with fifty or so cogs in it, exists in their nuclei. Each time the cell divides, the wheel moves forward one notch until, as the final notch nears, its rotation slows and finally stops. By the same token, a cancer cell may be immortal because its timer gets stuck at an early stage of development, so that it neither matures nor ages beyond that point.

A brief reminder here that life expectancy, the prediction of how long a given individual will live, is quite different from life span, the maximum number of years any member of that species has ever lived. Were medical science to conquer our two leading killer diseases, arteriosclerosis and cancer, these advances would add an estimated ten to sixteen years to the average person's longevity; it would not, however, extend his maximal life span of around one hundred and ten years. And so we age by increments, the fast-dividing cells of our skin, gastrointestinal tract, and blood going first and our slower-dividing cells, such as those of the liver and kidney, later.

If the Moorhead-Hayflick hypothesis is correct and, barring accident or fatal disease, the working of a biological clock determines the length of our stay on this planet, a logical question is: What makes this clock run down? There is ample evidence that, as a cell ages, its ability to adapt successfully to change declines. Do the myriad environmental insults our cells have to contend with gum up their clocks? Or might it be something more sinister, such as the action of a death gene?

For years—and perhaps still to some extent—the majority of gerontologists favored the view that the rate of accumulation of intracellular errors underlies the aging process, and that some catastrophic event, such as the loss of an essential gene, finishes a cell off. To put it another way: most cells are mini-factories whose genetic machinery, if not repaired promptly or replaced periodically, eventually breaks down. Studies have, in fact, shown that the ability of mammalian cells, including those of man, to repair the DNA-damaging effects of prolonged exposure to ultraviolet light correlates well with their life span. Thus, human beings repair their DNA at almost twice the rate of chimpanzees, who, on average, live half as long. Moreover, this repair capacity declines sharply as cells approach their reproductive limits.

The wear-and-tear theory likewise fits well with those age-related changes known to occur in the thymus gland and the brain. Weighing between 200 and 250 grams in infants at birth, the thymus maintains itself at this weight until around age fifteen, when, for some unknown reason, it begins to shrink—to 30–40 grams by age forty-five and a scant 3 grams by age sixty. Serum levels of various thymic hormones decline concomitantly until they become virtually undetectable. Though it is certain that the thymus plays a crucial role in immune system development in early life, since neither mice nor humans born without one survive long outside a protective environment, it is less clear how important its services are to adults. Many immunologists believe that loss of thymic hormones severely impairs the efficiency of the body's immune surveillance system. In their view, the progressive involution and disappearance of the thymus gland opens the gates to a multitude of disease-causing viruses and degenerative processes, and they may well be right—though personally I don't much care for this theory. If thymic hormones are so essential to good health, one would expect that the products of other glands, such as the pituitary or adrenals, would take over in their absence—as appears to be the case in those of my patients who have undergone surgical removal of the thymus for the treatment of the neuromuscular disease myasthenia gravis.

As for the brain, that mysterious organ "that obscures the secrets of its functioning even from itself," it may well control the "death clock" as it does almost everything else, including the timing of the menopause. For centuries, physicians tended to believe that a female was born with a limited number of eggs in her ovaries and once she hatched the last of them menopause ensued. But recent studies have exonerated the ovaries and placed the blame squarely on the brain. What brings on menopause is the failure of certain brain chemicals to trigger the cyclical release of the pituitary hormones that cause an egg to ripen and pop each month. And this programmed failure might not be an isolated instance of the brain's treachery. A cutoff of growth-stimulating substances to the heart, lungs, and other vital organs may well be what ages and ends us.

So, taken together, there is abundant evidence to support the wear-and-tear theory of cellular senescence. But there are major problems with it as well.

In a series of experiments begun in 1963, Wistar Institute scientists Jensen, Girardi, and Koprowski showed that infection of senescent WI-38 cells with the monkey tumor virus SV-40 stimulated most of them to begin proliferating again for a time, and some indefinitely—not the sort of comeback one might expect from supposedly worn-out, error-riddled cells.

An equally dramatic revitalization response was reported by C. W. Daniel and co-workers at the University of California, Santa Cruz, in 1984. When senescent mouse breast tissue cells were exposed to small, sublethal quantities of the toxin derived from the cholera bacillus—the same toxin that produces profuse and often fatal diarrhea in humans—they began to proliferate and form breast buds again.

Similarly, cell-fusion experiments by several scientists have shown that when cancer cells are fused with senescent cells, the latter's reproductive ability is restored for a limited time.

What these studies suggest is that senescence, rather than resulting from DNA damage, may actually be due to the switching off of some cell-replicating triggering mechanism that tumor viruses, or other substances, can turn back on.

An even more intriguing finding is that when senescent and young cells are fused, the young ones stop replicating. Moreover, as shown by C. K. Lumpkin of Baylor University, instead of cementing the two together, one merely has to inject the RNA from senescent cells into young cells to get the same growth-inhibiting effects. The oldsters were somehow signaling the youngsters to quit reproducing.

Before getting into the nature of such a strong signal and how it might act, let's look briefly at the ways cells communicate with one another. By analogy, they do so by radio, telephone, and messenger service. The most sophisticated of these is the *endocrine* system, whereby bulletins from one group of cells, such as those in the adrenal gland, are broadcast over long distances to many. Cells in closer proximity to one another usually use the *paracrine* system, which transmits signals via direct linkups or channels. The oldest and most limited is the *autocrine* system, in which a protein scout journeys outside the cell to look around a bit, then plugs back into a receptor on the cell's membrane and, through a series of relays, lets the nucleus know whether conditions are right for it to divide.

Primitive though it might be, the autocrine system is the only one available to the burgeoning embryo, which lacks a blood circulation and a nerve network, and is the main route by which cancer cells communicate with one another.

What do cells talk about? Probably the same sort of things we usually do: food, sex, and what's going on in the neighborhood.

There are several distinct stages in the life cycle of a cell. The first of these, the G_0 phase, immediately follows cell division and is essentially a rest period of variable duration until the beginning of the next go-round. The cell then enters the G_1 phase, where, depending on local conditions, such as the food and space available, and the influence of outside forces, it is directed along one of three pathways. It can go back into a proliferative mode, in which case it proceeds rapidly through the phases of DNA synthesis and mitotic cell division; it can undergo nonterminal differentiation, which suspends its reproductive cycle temporarily in order to equip it for some specialized function; or it can undergo terminal differentiation, never to give birth again.

Among the external forces acting on a cell in the G_1 phase is a family of proteins collectively called growth factors that, when local conditions permit, drive it toward proliferation. By fitting into cell membrane receptors that precisely match their configuration, such autocrine system messengers as interleukin-2 trigger a series of chemical reactions that, if strong enough, can reach the receiving station in the nucleus that initiates cell division.

Because the interaction between growth factor and receptor plays such a commanding role in cell behavior, much of the current research in both cancer and gerontology centers on it. When, in key-in-lock fashion, some growth factors and receptors combine, they form a functional unit that includes an enzyme called a protein kinase. Normally, this type of enzyme catalyzes the transfer of a phosphorus atom from a donor to an acceptor molecule, but this one doesn't need a donor; seemingly it plucks a phosphorus atom out of thin air instead. This three-part complex is then drawn into the cell, where its enzyme portion passes the phosphorus atom on to a series of high-energy compounds called G proteins that give it a strong push in the direction of the nucleus.

With this process in mind, let's examine what might happen when a robust young cell comes under the inhibiting influence of an older one. As shown by Baylor University and Wistar Institute investigators through different means, the transfer of an anti-proliferative protein, or else the RNA coding for it, prevents the receptor complex from responding to the growth factor signal that activates it to propel a phosphorus atom on its way to the nucleus to trigger cell division. And when cells lose their ability to replicate periodically, they engorge with waste products, spring leaks in their outer membrane, and disintegrate. In short, a death gene is at work.

In sharp contrast, the primary action of many of the known oncogenes is on these same receptors. Either by flooding a particular type of receptor with growth factors or by jamming it in the "open" configuration, such runaway genes force the cell into continuous and eventually cancerous divisions. Which is one of the reasons why Dr. Cristofalo believes cancer may well be the antithesis of aging.

The question that has long haunted Cristofalo, as it must all gerontologists, is: Does a way, or ways, exist to rejuvenate a cell without the risk of turning it cancerous? And the very tentative answer he has come up with is yes.

What Cristofalo's group has found is that one age-related glitch in a cell's reproductive triggering mechanism develops in that portion of its membrane receptor acting as an enzyme. Although the exact nature of the defect is unknown, the fact that it can be overridden for a time through infection of the cell with tumor viruses, or other lab manipulations, proves that it is reversible—or, in scientific terms, functional rather than structural—but how to safely reverse it?

Thanks, in part, to the monoclonal antibodies created for them by Wistar Institute colleague Barbara Knowles, Cristofalo's team was able to isolate the membrane receptor to a powerful growth stimulus, epidermal growth factor, and test its function under a variety of circumstances. Two major findings emerged from these delicate experiments: (1) For old cells, the receptor's phosphorus-binding enzyme worked well when the entire membrane was stripped from the cell, but paradoxically, not at all when the receptor complex was separated from the membrane.

(2) The addition of certain membrane fragments from young cells to the purified receptor preparations of old cells promptly restored their enzymatic activity. From these findings, Cristofalo concluded that there are both stimulatory and inhibitory factors in and around this growth factor receptor, and the existence of these fine-tuning elements explains how anti-proliferative proteins from senescent cells and membrane fragments from young ones can act in opposite ways.

Dr. Cristofalo readily admits that any conclusions drawn from these experiments are at best preliminary and that, apart from their potential benefit to senescent cells in culture dishes, he has no idea what such membrane fragments, with their intimations of immortality, have to offer. Cautious researcher that he is, he declines to speculate on the possibility that restoration of the membrane receptivity and replicative ability of certain senescent cells might be enough to revitalize an entire tissue, organ, or organism; however, on the basis of the accepted findings of Hayflick that the maximal number of times a cell can divide in culture is proportional to the life span of the species from which it was derived, there is reason to hope that this will prove to be the case. In the meantime, what is the age-fearing person to do? Are there any scientifically sound stopgap measures?

Actually there is one: diet—though not just any diet will do. The intake of calories has to be sufficiently low to deceive the brain into thinking: starvation. In response to this seeming emergency, or so the theory goes, the brain releases a variety of survival hormones that, among their other functions, slow the rate at which many types of cells replicate. While the maximal number of doublings these cells can undergo remains the same, they simply take longer to run out the string. Laboratory studies of young rats have, in fact, shown that drastic caloric restriction can extend their customary life span of around three years by as much as 50 percent.

And still available to those who choose not to live out their days in perpetual hunger are the old-fashioned ways of achieving immortality: do great deeds, create great works, or have children.

14

A STORY that I've long found fascinating and told many times involves the late Chinese premier Zhou Enlai and his adopted daughter, Li Bing. Upon graduating from medical school, she announced to him: "Father, I have decided to devote my life to curing cancer," thinking the lofty goal would please him. To her surprise, he turned to her with scorn. "How can you make such a ridiculous statement," he demanded, "when you know so little about your enemy, his strengths and weaknesses, his tactics, how he deploys his troops"—using military analogy after analogy to point out how ill-prepared she was to undertake such a mission. Yet, rather than being discouraged, Li Bing saw wisdom in the admonition. Over the next several years, she took the lead in organizing close to a million of the so-called barefoot physicians of China to reconnoiter and record the enemy's presence throughout the entire country. From 1972 through 1975, these teams obtained near-complete data on the causes of death among 800 million of the inhabitants of the People's Republic and published them in an atlas of cancer mortality that one needs merely to glance through to grasp its extraordinary significance.

Cancer ranks second only to respiratory diseases as a cause

of death in China, accounting for 11.3 percent of the deaths among males and 8.8 percent among females. It is the leading killer of those over the age of fifty-four. Though these statistics are of some interest, what the survey revealed that is both startling and mystifying is dense clusters of a particular type of cancer in certain geographical regions, strongly suggesting that some local environmental factor is at work. For instance, stomach cancer, the most common malignancy in China, is much more prevalent along the Sino-Soviet border and in the northeastern coastal areas, esophageal cancer in the northern sections of Szechuan province and portions of central China, and colon cancer in the delta regions along the Changjian River. In contrast, the occurrence of liver cancer is much more frequent along the southeastern coast and, as might be expected, lung cancer in urban centers, such as Beijing, where industrial pollution is a major problem.

Explanations for these clusters range from the common practice, especially in areas lacking electrical refrigeration, of preserving foods by smoking or pickling them, to viral and parasitic infections indigenous to the regions. Or they may possibly arise from a combination of local factors similar to those discovered by University of Glasgow researcher William Jarrett to be associated with a high incidence of gastrointestinal cancers in cattle. In a brilliant piece of detective work, the veterinary medicine team headed by Jarrett found that, though neither infection by a subtype of papilloma virus—so called because it produces widespread papillomas, or warts, some of which evolve into skin cancers—alone nor the ingestion of bracken fern, a plant known to contain carcinogens, could account for the malignancies, the two factors together can.

The monumental survey completed by Li Bing and her

Ministry of Health colleagues has stimulated many other groups to undertake long-term studies to determine what variables correlate with susceptibility to certain types of cancer. I'm not exactly sure how they did it, or how well their estimate will hold up, but in 1982 a committee of our National Academy of Sciences concluded that dietary factors exert a major influence on approximately 60 percent of the cancers in American women and 40 percent of those in men. By and large, Wistar Institute associate director David Kritchevsky agrees, and much of his research centers on the effects of dietary composition on the development of chemically induced cancers in experimental animals. Out of his work have come several important discoveries, the most recent concerning dietary fiber.

Based mostly on indirect evidence, the National Cancer Institute's food advisory panel has recommended that we all eat more fiber—and as a television actor in a breakfast cereal commercial says, "Who am I to argue with the National Cancer Institute!" Others, however, including many nutritional experts, are not so reticent; they contend that the evidence in favor of fiber protecting against colon cancer is not yet strong enough to justify such a major change. In fact, Wistar Institute scientists David Klurfeld and David Kritchevsky have recently confirmed the preliminary finding of Michigan State University food scientists that, in some forms, fiber may actually promote colon cancer. Knowing that "fiber" is merely a generic term encompassing a wide variety of substances, Drs. Klurfeld and Kritchevsky compared the effects of several soluble versus insoluble forms of fiber on the development of colon cancers in rats exposed to a powerful chemical carcinogen. Their surprising results reveal that, while soluble fibers, such as the pectins derived from citrus fruits, reduced the risk of tumor formation, insoluble fibers, such as powdered cellulose, increased it. Although cautious about the implications of his team's findings, Dr. Kritchevsky feels that the blanket prescription for more fiber in our diets should be reconsidered and, if rewritten at all, done in more specific terms.

Since certain common malignancies, such as breast and uterine cancers, appear related to body-fat content, afflicting the obese with approximately twice the frequency that they do the lean, some nutritionists recommend a reduced dietary fat intake to protect against these cancers. But in his animal models, Dr. Kritchevsky has shown that the restriction of total caloric intake, rather than fat, is much more effective in this regard. For example, rats fed 40 percent fewer calories a day, even when the special food they received was rich in fat, developed far fewer and smaller breast and colon cancers than did rats fed a low-fat, normal caloric diet. Hence, under the conditions of this experiment, changes in body fat and protein stores appear to exert a greater influence on the growth of these cancers than does the amount of fat ingested. From these studies, Dr. Kritchevsky concludes that, for the prevention of cancer, if not in its management, the composition of one's diet is less important than total daily caloric intake.

Although medical science has gleaned much about the relationship between food and cancer in recent years, very little of it is encouraging. To give an old adage an unoriginal new twist: almost everything truly enjoyable in life is not only expensive, immoral, or fattening but probably carcinogenic as well. Since dietary fat has been linked to breast and prostate cancers, excessive salt and carbohydrate to stomach cancer, and meat to colon cancer, what is left for one to eat? I scarcely need mention the potential, if not actual, hazards we face from the three thousand or so food additives we all ingest at one time or another, along with the countless chemicals that get into our food sources accidentally. On the bright side, however, Dr. Kritchevsky's animal experiments suggest that even a 10 percent reduction in total calories can reduce the incidence of certain cancers by as much as 40 percent.

The reason why most cancers take many years to develop is that they must pass through the stages of initiation, promotion, and progression before attaining sufficient size to become clinically apparent. The first of these stages may be brief, but traumatic: a single exposure to a cancer-causing agent that damages a small, strategically placed piece of DNA. Given time, the cell can usually mobilize the necessary enzymes to repair the damage. But should it divide before such repairs are completed, the abnormality will likely be passed on to its progeny and become genetically fixed. What, in many instances, forces the damaged cell to divide is a physical or chemical agent called a tumor promoter. By itself, a promoter is usually not capable of turning a cell cancerous, but once an initiating event has occurred, it can act as a malignant adviser, urging the cell to break free of the forces that keep it from growing and go its own independent way.

Once a cell, be it of the fast-dividing types in the skin and gut or the slow-dividing types in the liver and lung, is exposed to first a DNA-damaging carcinogen such as a hydrocarbon, and then a tumor promoter, it develops a memory of sorts; a further encounter with the promoter, even many months later, can drive the cell wild.

The significance that this finding holds for cancer researchers lies in the fact that a diverse group of drugs, chemicals, food additives, and building materials can act as tumor promoters—

among them alcohol, phenobarbital, dioxin, saccharin, asbestos, and tobacco. In certain combinations these substances can prove deadly indeed; for example, the chance of asbestos workers who are also heavy smokers contracting lung cancer is 900 times greater than that of nonsmokers not exposed to asbestos fibers in their daily occupations.

Taking tumor promoters as their special point of attack, Dr. Leila Diamond and her Wistar Institute colleague Dr. Thomas O'Brien are trying to discover exactly how some of them work. One approach has been to compare a series of compounds with similar, though not identical, chemical structures to determine which of these endows it with tumor-promoting properties. From these studies, researchers such as Dr. Diamond have formulated the theory that, to exert this type of action, a substance has to fit into a cell membrane receptor meant for a hormone or growth factor.

Once again, a cell is pretty much a closed and tightly protected piece of protoplasm; to get inside, a molecule must usually pass through the special portal made for it. In the case of a growth factor, what happens next is a series of chemical reactions to activate the enzyme on the inner, cytoplasmic side of the receptor. This, in turn, starts a cell-dividing signal on its way to the nucleus. In contrast, a tumor promoter is capable not only of slipping into a membrane receptor but of altering it in such a way that several of the usual chemical steps are circumvented and a powerful growth signal generated. The substances most commonly employed to study this system are the phorbol esters. Many of these compounds, such as croton oil, are extracts of plants that for centuries have been used as herbal remedies in China and may possibly account for some of the unusual cancer clusters that occur there. Though the pathway by which phorbol esters stimulate cell proliferation is not yet clear, recent evidence suggests that the end result is activation of certain nuclear oncogenes, such as MYC—which, as we have already learned, is a dangerous gene to get stirred up.

Should it turn out that most tumor promoters operate through a common mechanism, or a limited series of them, it then becomes feasible to devise safe ways to block their actions. This probably wouldn't be enough to destroy cancer colonies that are already established, but might well prevent new ones from forming. Clinical studies are underway to test whether certain vitamins that block the action of tumor promoters on cells in the test tube are equally effective when administered in large doses to high-risk groups, such as coal miners and asbestos workers. I would, however, caution those who, rather than wait for the results of such studies, want to get a head start that they might be putting themselves at greater risk of dying from vitamin A or D poisoning than from cancer. Better to consume more fruits and vegetables and wash them down with something other than an alcoholic beverage.

Once a cancer cell has successfully navigated the hazards of early life and motherhood, its next major obstacle becomes finding the room to keep multiplying—for that is really all it knows how to do. Since its neighbors, especially in densely packed tissues, are unlikely to prove accommodating in this respect, the cancer clone must invade, subvert, destroy, and cannibalize them in order to gain what Hitler once called *Lebensraum*, or living space, without limits or end.

Dr. Clayton Buck, who joined the Wistar faculty in 1975 and was recently appointed its Director of Scientific Development, researches the differences in the way normal and malignant cells adhere to one another. Tall, lank, and bespectacled, a wool plaid shirt and faded jeans under his white lab coat, Dr. Buck and Wistar Institute professor Leonard Warren were the first to describe the characteristic changes in carbohydrate content that occur in tumor cells.

Most cells are not surrounded by air-filled space; they are immersed in a gelatinous substance called the extracellular matrix. A dense network of cables crisscross this semi-liquid material to anchor its major structures in place. A second series of microfilaments and microfibrils radiate from each cell's nucleus to all points on its membrane. For a cell to maintain its proper shape and position, or to move elsewhere along grooves in the extracellular matrix, these two sets of cables must interconnect in a fibrous mesh of proteins, termed the basement membrane, replete with receptors and adhesion molecules.

To invade a neighbor, a cancer cell must have the chemical capability to do two things: break the cables that bind it in place and chew its way through its victim's basement membrane by releasing various protein and connective-tissue-dissolving enzymes.

Normally, when two cells bump into one another they form glycoprotein structures called *adhesion plaques* to halt their forward progress and to act as "checkpoints" for molecules passing between them. Most cancer cells, however, appear incapable of either making or binding to these moorings, most likely because the surface actions of certain of their oncogenes prevent it. This abnormality is not absolute, however; when, through various test tube manipulations, the missing protein cables and connectors are provided, cancer-cell migrations usually cease and they behave themselves.

From studies of both normal and tumor-virus-infested chick

cells, Buck's team has isolated a family of basement membrane molecules, the integrins, that serve to connect cables from inside and outside the cell. Moreover, in cells transformed by the Rous Sarcoma Virus, the chemical structure of these molecules is so altered that they can no longer bind these cables tightly—leading Dr. Buck to conclude that the integrins represent one of the targets for the oncogene carried by this extremely potent tumor virus, and possibly others as well.

Recent experiments have demonstrated that the ability of skin-cancer cells to invade mouse tissue can be modulated by adding or removing certain adhesion molecules from the culture in which these cells are grown before being injected into the animal. Eventually, Dr. Buck's team hopes to discover safe, effective ways to block the spread of cancer cells not only in the culture dish and experimental animal but in the patient.

Another strategy to halt a cancer's growth by choking off its blood supply is being developed by Wistar Institute professor Elliot Levine. Like any rebel force in desperate need of sustenance but no longer able to obtain it solely off the land, a cancer must open new lines of supply in the form of blood capillaries. For this, it needs to recruit the endothelial cells that constitute the blood vessel's scaffolding and inner lining. It is this layer that represents the vessel's business end, permitting or preventing the passage of nutrients, oxygen transporters, cholesterol, and cancer cells into the tissues around it. Accordingly, our endothelial cells come in plentiful supply. Dr. Levine estimates that, if extracted in toto and spread out, an average-sized person's endothelium would weigh around four pounds and cover an area of approximately two acres.

In 1971, Harvard Medical School researcher Dr. Judah Folkman proposed the theory that many cancers produce chemical substances to stimulate the capillary growth process, and since then several such "tumor angiogenic factors" have been isolated and identified. With this came the hope that safe ways might be devised to block the actions of these factors, deprive the cancer of new blood vessels, and, if not starve it to death, at least keep its growth in check. But all attempts so far to find a class or combination of inhibitors that act as effectively in the patient as in the test tube have failed. What, until recently, impeded progress was lack of a reliable method of growing human endothelial cells in culture to serve as a test model. But now, after years of intensive effort, Dr. Levine's team has finally found the right mix of ingredients to get such human cells to grow abundantly and for long periods in the laboratory. Moreover, the properties of these homegrown cells closely mimic those of their bodily counterparts: they proliferate in response to the appropriate growth factors; they mature, specialize, senesce, and cease replicating in accordance with the usual timetable; some of them, when exposed to a carcinogen, are even transformed into blood-vessel tumors. A particular hope of Dr. Levine's is to learn how a cancer attracts and makes room for the migratory endothelial cells needed to form the capillaries that the cancer cannot make on its own. With this system in place at The Wistar Institute and elsewhere, chances are good that scientists will soon identify all the major chemical attractants *and* repellents of endothelial cells and find exciting clinical uses for them as well.

Possibly the most difficult of the challenges currently confronting Wistar Institute scientists is the one Dr. Giovanni Rovera has taken on: how to convert leukemic cells into mature normal cells by providing them with the necessary genetic instructions. In his late forties, "Gianni" Rovera is medium tall, slim, and distinguished-looking. By training a hematologist and pathologist, as well as a molecular biologist, he is married to Wistar Institute associate professor Daniela Santoli, whose research into T-cell function is closely allied to his own. In Hilary Koprowski's opinion, "Gianni Rovera probably has the broadest scientific knowledge of anyone at the Institute." Yet, as Rovera is the first to admit, he will likely need every bit of it to find the puzzle pieces that have eluded him for the past decade.

In common with many cancers, the growth patterns of leukemic cells reflect two interrelated sets of genetic abnormalities: the first, reducing or eliminating their need for external growth factors; the second, overriding those forces that would put the young cell on the road to maturity and specialization and switching it to the one for perpetual motherhood instead.

That these genetic changes can conceivably be overcome is suggested by the finding that, either spontaneously or through the addition of certain chemicals, a small proportion of leukemic cells do mature and "differentiate" in the sense that they develop some special function. With the advent of recombinant DNA technology, and the availability of a widening assortment of potent cell-differentiation factors, such as certain of the interleukins, interferons, and colony-stimulating factors, Rovera's team, along with others in the field, had high hopes that, with the right combination, they could perform the "conversion" trick. But so far none has with any consistency. In nearly all the battles fought for control of the cell cycle, the oncogene forces have won out over the differentiation factors. It has therefore become clear to Dr. Rovera's team that before a leukemic cell can be tamed, its oncogenes have to be dealt with—and in approximately half the leukemic cell lines they have studied thus far, the principal one appears to be RAS (probably in conjunction with MYC, which is less accessible to attack).

In collaboration with Drs. Prem Reddy, Carlo Croce, and others, Dr. Rovera is currently trying to devise a specific inhibitor of the RAS oncogene's protein and engineer its entry into the cell. Should the effort succeed, it might prove all that is necessary to convert leukemic cells into normal ones; if not, the Wistar group will have to figure out what more needs to be done to convince such cells to abandon their wicked ways and follow the path of the straight and narrow.

15

IN MID-JULY, I met Jim and Tinka Federline in the rose marble lobby of my favorite Washington, D.C., hotel, the J. W. Marriott on Pennsylvania Avenue. Jim was handsomely decked out in a lightweight sports jacket and gabardine slacks. A tan he had acquired during his family's recent stay at their beach house masked some of his skin pallor, and though perhaps a little thinner and more hesitant in his movements, he looked not too much different from the last time I had seen him. Tinka, as always, was energetic and cheerful, the very personification of the positive thinker. At the time I couldn't help wondering how much of this bravura was genuine and how much feigned to bolster Jim's spirits, but as I've come to know Tinka better, I've learned that this is her way.

For the next several hours we sat in the decorous lounge off the hotel lobby and talked. During our last visit together, I had broached my intention of writing a book about cancer research that would include, and possibly feature, Jim's case, and the Federlines offered their full cooperation. A question they had not been prepared to answer, however, was whether they preferred fictitious names to their real ones when I wrote about them. The way we had left it was that they would talk to their two daughters to see how they felt about having the family's ordeal made public.

Now, with pride in her voice, Tinka told me that their girls had decided they wanted the Federline name used "so that they might someday show this book to their future children to let them know about the experience they had all gone through as a family." In her very next breath, however, Tinka made it clear that they had no intention of letting Jim's story end sadly or any time soon. One of their closest friends had repeatedly advised Jim to give up being experimented on and "let Nature take her course."

"Let Nature take her course with somebody else!" Tinka said vehemently. "We're going for a cure."

With fond memories of their Nantes, France, experience and the temporary gains Jim had made following his treatment there, the Federlines asked if any of The Wistar Institute researchers I'd spent the past week with had come up with anything of similar promise that Jim might try.

Having already asked the same question of Zenon Steplewski, Dorothee Herlyn, and others at the Institute working on new forms of treatment, I knew that the answer for the moment was no. On the basis of test tube experiments showing that gamma-interferon markedly enhanced the ability of immune system cells to destroy cancer cells, The Wistar Institute had entered into a collaborative study with Philadelphia's Fox Chase Cancer Center to test the effectiveness of a combination of monoclonal antibodies and gamma-interferon infusions in a series of gastrointestinal cancer victims. But the results obtained in the first twenty-seven patients so treated were disappointing; even when massive quantities of gamma-interferon were administered, the patients—all with far-advanced cancers—responded no better than those given antibodies alone. Having previously discussed with Jim the possibility of his receiving such combined therapy, I now had to tell him the chances of his benefiting were so slim that it didn't warrant the risk.

Jim nodded, having already gathered this from phone conversations with Zenon Steplewski, and went on to ask about another new form of immunotherapy. Six months earlier, in an article in the *New England Journal of Medicine,* Dr. Steven Rosenberg and twelve of his National Cancer Institute co-workers had reported the encouraging, if mostly temporary, responses of twenty-five metastatic cancer victims to a combination of interleukin-2 and a special type of lymphocyte, called an LAK cell, derived from each patient individually and grown in large numbers outside the body before being reinfused into him. All in all, one patient had responded completely and ten others—three with colon cancer—"partially," in that their tumor masses shrank by more than 50 percent. In an unusual display of "medico-media" hoopla, Dr. Rosenberg's findings made the front page of virtually every major newspaper in the country, and to add further legitimacy to this new form of treatment, NCI director Dr. Vincent DeVita, in the *Newsweek* magazine cover story that appeared shortly afterward, proclaimed it "the most interesting and exciting biological therapy we have seen so far." Despite warnings by Dr. Rosenberg and others that the value of interleukin-2 treatment was far from proven and its side effects severe, the National Cancer Institute subsequently received thousands of phone calls and letters from desperate patients inquiring where they might obtain such treatment. With his local oncologist acting as intermediary, Jim Federline was among them, only to run up against an obstacle many cancer victims have to contend with as they struggle to stay alive—that of "protocol" restrictions. To keep variables at a minimum and so help ensure reliable results, those in charge of patient selection for such therapeutic trials often have no choice but to consider only candidates who, except for failed surgery, have not otherwise been treated. Because Jim had already received two experimental therapies—Tumor Necrosis Factor and monoclonal antibodies—he was ruled ineligible for any of the NCI-sponsored interleukin-2 trials. Was it even worth pursuing, Jim asked me, and if so, where?

I had no good answer to the first part of his question. The interleukin-2 plus LAK cell treatment of cancer was simply too new, too controversial. I did, however, have an answer of sorts to the second part.

In a radical and much criticized departure from the customary way clinical research is conducted, Dr. Robert Oldham, a former NCI program director, had not only founded his own cancer treatment center in Franklin, Tennessee, but instituted a new way—"fee for service research"—to fund it. Instead of depending on the vagaries of federal or pharmaceutical company grants to subsidize his clinical studies, Dr. Oldham charges each of his patients in advance a sum ranging from nine to thirty-five thousand dollars for the experimental therapies they would then receive.

Jim had heard about Dr. Oldham and asked my opinion of the man and his Biological Therapy Institute. Except for the fact that my close friend Dr. Kenneth Foon had worked with Robert Oldham at the National Cancer Institute and knew him to be a competent investigator and a leading authority on the interferons, I had little else to tell Jim—though I did think phoning Dr. Oldham to find out what his institute might have to offer would be worthwhile.

With that bit of business out of the way, I then asked Tinka and Jim a series of questions about how they had met, married, raised a family, and made such a great success of the plumbing and heating firm begun by James Federline, Sr. I could see the pain in Jim's eyes as he described his father's slow, agonizing death from lung cancer and why, after watching him suffer horrendous side effects from aggressive chemotherapy, Jim was so adamantly opposed to receiving it himself.

At the end of the afternoon, I accepted the Federlines' invitation to have dinner with them. One of their favorite Washington restaurants was six or seven blocks away and, even though the July day was sweltering and I was more than a little concerned how Jim might handle the exertion, we walked to it.

I knew that, because of the fluid compressing his stomach and the 3,000 calories of intravenous nourishment Tinka was feeding him daily through the indwelling venous catheter in his neck, Jim's appetite was virtually nil. Yet he ordered the same substantial meal that Tinka and I did, and by picking at it steadily he managed to consume around a third of it.

As the dinner progressed and we discussed various research leads, I found myself identifying more and more with Jim, aware and grateful that his plight was not mine, yet wondering all the same whether I had it within me to bear its burden as well as he. A doctor friend who had served as medical officer to a front-line infantry regiment in the Korean War once explained to me the subtle difference between courage and bravery—the gist of his argument being that one displays courage by making the best of a difficult or dangerous situation; bravery by risking great personal harm, even death, in an effort to change it.

Clearly, I concluded, Jim Federline was acting bravely, and quite likely because of it, his immune system was able to keep up its tenacious struggle against an enemy constantly threatening to overwhelm it.

Although the extent to which a human being's emotions can affect his bodily defenses remains unknown, it's my hunch that, if the brain holds the same sway over the immune system as it does over the endocrine system, the answer will turn out to be: a great deal. As an internist whose subspecialty is endocrinology, I have encountered over the years two striking examples of the influence psychological factors can exert over certain glandular functions—so striking, in fact, that they might be termed psychoendocrine spectaculars.

The first of these is the well-known, though still somewhat mysterious, condition called anorexia nervosa, a malady that primarily afflicts teenaged girls. Always a serious and sometimes a fatal disease, its existence is proof that the late Duchess of Windsor was wrong when she allegedly remarked that "one cannot be too rich or too thin." What physicians who treat such patients have long found remarkable is their triad of denials: though they may have lost so much weight that they are skeletal in appearance, they deny they're hungry, deny they're tired, and *even deny that they're thin*. Most had been on the plump side to begin with and as still another manifestation of their disease have developed a "fixed body image." On occasion, to prove that the way they perceive themselves in the mirror is grossly distorted, I have raised the patient's arm to her face and exclaimed, "Look! Look how thin this is!" The usual reply: "Yes, isn't it funny how I've lost weight only in my arms."

Though as yet inconclusive, biochemical studies of anorexia nervosa patients strongly suggest that their brains put out excessive amounts of one of the morphinelike family of substances known as the endorphins. In addition to their painkilling properties, these brain chemicals are capable of exerting a wide range of biological effects that include inhibiting the cyclical release of the pituitary hormones that induce menstruation, modulating the hunger drive, and producing a false sense of well-being. Discovery of the endorphins may lead not only to a better understanding of the peculiarities of this disease but to better treatment for it as well. And the more recent finding that specific endorphin receptors exist on white blood cells has stimulated immunologists to try to figure out exactly what sort of messages these brain chemicals carry.

The second type of psycho-endocrine spectacular that on rare occasions I have seen is called *pseudocyesis*, or "false pregnancy." In this condition a woman is so intent on bearing a child that her menstrual periods cease, her breasts engorge with milk, and her abdomen swells progressively. Yet pregnancy test after pregnancy test proves negative.

So, although uncertain as to the degree of their impact and unwilling to go overboard about them, I believe that emotional extremes can affect our immune defenses as profoundly as they do other bodily systems, most likely through changes in those brain centers, such as the hypothalamus, where mood and hormonal regulators coincide. It might even be supposed that at some very early stage in the development of certain diseases, this emotional factor can be critical in determining their future course. However, as I am in a relatively calm period of my life, this may represent more wishful thinking than sound scientific speculation on my part.

That the brain is capable of influencing immune system function may seem obvious, but for a long time it wasn't. Since T- and B-lymphocytes, when appropriately challenged, seem to respond as well in the test tube as in the body, many immunologists formerly believed that the immune system operated independently of the central nervous system. Nowadays, however, all agree that the trillions of cells that comprise the two systems are tightly interwoven. Some neuroscientists even credit macrophages and T-cells with the ability to feed information to the brain about cellular events, such as viral invasions, that it might otherwise be unable to detect. Certainly the rich network of nerve fibers that run from the midportions of the brain to such immune system depots as the bone marrow, spleen, thymus, and lymph nodes provides a direct linkup between the two systems. Transmissions via various brain chemicals, hormones, and lymphocyte products, such as the interleukins, are likewise frequent.

Among the brain's stimulators of immunological function, B-endorphin appears to exert a particularly potent effect on the type of immunocytes known as Natural Killer cells. Unlike the mainstays of the body's immune defenses, the T-cells and B-cells, which may take as long as two weeks to receive and respond to their marching orders, Natural Killer cells can recognize and destroy an enemy target within hours and so constitute a mobile strike force. Together with macrophages, they compose the body's first line of defense against cancer cells. Moreover, test tube experiments demonstrating that the addition of B-endorphin to a population of Natural Killer cells markedly increases their ability to kill diseased cells, leave little doubt that these cells thrive on this type of brain stimulation.

The converse also appears to be true. Studies show that the biochemical changes accompanying the clinical state of depression—especially that following the loss of a marriage partner—can severely impair immune defenses. In an attempt to quantify this phenomenon, two investigations of immunological function in bereaved individuals have been carried out, one by Australian researchers and the other by a team of psychiatrists from Mount Sinai Medical School in New York. Each used as its end point the test tube response of lymphocytes derived from these patients to the addition of chemical factors that normally stimulate their growth and activity. In the first of these studies, involving twenty-nine test subjects whose spouses had died two to six weeks earlier, Australian investigators found a marked reduction in the lymphocyte reactivity of these individuals when compared to that of a control group. In the Mount Sinai study of six men whose wives suffered and eventually died of metastatic breast cancer, serial measurements showed the same steep decline in lymphocyte function, which, on the average, reached a nadir one to two months after their wives' deaths and usually, though not invariably, returned to normal within six months.

What is new and noteworthy about these studies is that they provide an explanation for the well-known fact—recently confirmed by a 95,000-patient Swedish study—that widows and widowers suffer an abnormally high mortality rate during the year following their loss. They also carry a warning that I do not hesitate to broach to my bereaved patients: that unless they can find reasons and ways to overcome their grief fairly quickly, they increase the risk of dying themselves.

Most white blood cells have receptors not only for the endorphins but for several other chemicals the brain uses to transmit nerve impulses, such as serotonin. Many of the drugs physicians use to successfully treat depression work, at least in part, by correcting deficiencies of this chemical in certain pleasure-regulating areas of the brain.

Another way in which chronic depression makes life miserable for its sufferers—and their immune systems—is via a mechanism that increases the adrenal gland's output of the immunosuppressive hormone cortisone. Normally, through an elaborate arrangement known as a "feedback loop," cells in the hypothalamus, either on their own or on the advice of higher brain centers, decide whether or not the body needs more cortisone. Should the answer be yes, these cells send a signal to the pituitary to secrete adrenocorticotropic hormone (ACTH), which, as its name implies, has a special affinity for the outer layer of the adrenal gland. ACTH then stimulates the manufacture and release into the bloodstream of enough cortisone to convince the hypothalamus that the need has been met and to shut its stimulators off. Or that is the way the feedback system is supposed to work. For some still unknown reason, the brains of chronically depressed individuals continue to pump out ACTH long after the normal shutoff concentration of cortisone has been reached. And although one cannot live long without this essential blood-pressure- and blood-sugar-regulating adrenal hormone, one cannot live comfortably with a large surplus of it either. Many of the complaints, such as fitful sleep and frequent infections, that plague depressed patients are now ascribed to the excessive amounts of ACTH their brains put out.

Severe stress is no friend of the immune system either, especially the sort over which one can exert little if any control. For example, in an experiment designed to test the effects of controllable versus uncontrollable pain and anxiety on an animal's susceptibility to cancer, researchers injected live cancer cells into a group of rats and then put them in chambers wired to give electrical shocks. Half the rodents were afforded the chance to terminate the shocks temporarily through an escape route, and half were not. The result: three-quarters of those rats given a way to elude the shocks rejected the cancer cells, compared to only one-quarter of those having no control over the shocks.

How does emotional turmoil adversely affect one's immune defenses? Doubtless cortisone plays a role, as does the stress hormone supreme, epinephrine—more commonly known as adrenaline. Because a creature engaged in "fight or flight" needs both a blood pressure to support it in the upright position and plenty of sugar to burn, epinephrine is capable of boosting the adrenal glands' output of cortisone in two ways: by stimulating the pituitary gland's release of ACTH and by increasing the adrenals' concentration of certain cortisone-making enzymes. In a "one hand washes the other" sort of arrangement, cortisone then returns the favor by activating another set of adrenal gland enzymes to manufacture more epinephrine, and unless higher brain centers intervene to halt the process at this point, a vicious cycle ensues.

So one way in which emotional factors can cripple one's immune system is through the adrenal hormones, epinephrine and cortisone. Another probably comes about through deficiencies in certain brain chemical stimulators at the cell membrane level. And a third involves foul-ups in intercellular communications that the immune system brings on itself.

Although the traffic in messages seems to flow more heavily from the central nervous system down, certain immunocytes, by means of such chemical signals as the interleukins and interferons, do inform and advise the brain on several matters. If, for instance, the immune system regulators at the site of a viral invasion determine that not only has the threat been adequately contained but its killer cells are getting a little out of hand, they can shoot a signal to the adrenal glands, via the hypothalamus and pituitary, to send more cortisone to the scene to call off the fight. Or, should the brain dispatchers be preoccupied with other problems at the moment and too slow to respond, the local immune system regulators can assign a contingent of the white blood cells under their command to make their own ACTH and cortisone. The thymus gland is also integrated into the system to play a dual role: its hormones trigger the release of ACTH and cortisone in one direction and the growth and development of T-cells in the other. Even the sex hormones, estrogen and testosterone, get into the act through the effects they exert directly on the thymus and indirectly on certain types of T-cells. Thus, the nervous, endocrine, and immune systems interrelate at several levels and, just as with military units in battle, must communicate clearly with one another to avoid potentially ruinous mistakes. A lapse or laxity in immune system surveillance can enable a cancer to take root; a misdirected or excessive response—say, to a harmless virus, drug, or wind-borne pollen—can result in a severe allergy or even an autoimmune disease, such as ulcerative colitis. In the most succinct and eloquent summary of the relationship between emotions and immunologically regulated diseases I have read, the famous pathologist William Boyd wrote: "The sorrow that hath no vent in tears may make other organs weep!"

Knowing what I do of the life stories of Norman Arnold and Jim Federline, a question I've often pondered is whether their largely unexpressed and unresolved grief over the deaths of their fathers could have triggered their pancreatic cancers. The answer, I believe, is: probably not. I regard the interplay between emotions and cancer in much the same way as I do that between a record player and a phonograph record. Though the knobs on the set can modulate the sound coming from it, without a record, a music source, on the turntable, they cannot create it. Nor do I believe—as do those who advocate the "guided imagery" method of combating cancer by visualizing the destruction of cancer cells in the mind—that psychic factors, however powerful, are enough to reverse the course of the disease once it has entered a rapidly progressive phase.

The day may well arrive when we are sufficiently versed in the ways of brain and body to practice such self-healing successfully. We do, after all, have within us, and in purer form than any pharmaceutical company can supply, all the disease fighters—the cortisones and interferons and antibodies—we need; we simply haven't figured out how to summon them at will when we are critically ill. Maybe, at some unconscious level, our brains do know, but won't cooperate. If so, and if the techniques that medical scientists are currently developing to test the mechanisms by which the mind influences the body work well enough, I'm hopeful that many of these secrets will soon be revealed. With a little luck, we may even learn how, though probably not why, our cells have been preprogrammed to die.

16

AT AROUND the same time that Jim Federline journeyed to Tennessee to consult with Dr. Robert Oldham in late July, the following letter appeared in the Flint *Journal* under the caption AFRAID FOR CANCER VICTIMS. I include it now because to my mind the two events share a similar theme.

> I recently lost my father to cancer. I hurt and I am angry. I am angry because Oral Roberts is asking for millions of dollars. I am angry because there is a frantic scramble to find a cure for AIDS.
>
> I am angry because we have not found a cure for a disease that has been killing people unmercifully for years and years. I am angry because there is not a pain medication available that will allow cancer victims to live pain-free but be coherent enough that they can have "quality" time with their families.
>
> I am angry and I am afraid. I am afraid for the people I love. I am afraid for the people you love. One out of five Americans with cancer will die of it this year. Why?
>
> Paula L. Meredith-Faris
> Burton, Michigan

An informed reply to Miss Meredith-Faris's question depends largely on how one answers another: Given the resources—the funding, personnel, public awareness campaigns—that a panel of the experts deems necessary, can we do much more to cure cancer than we are already doing? Why even consider such a crash program when over twice as many Americans die of arteriosclerotic heart disease each year and national polls indicate that more people are afraid of contracting AIDS than cancer?

One reason concerns hope. Having talked to enough patients and personal friends about how they felt on learning they had a life-threatening illness, I now have some inkling of the special numbness, horror, and sense of self-betrayal that comes with the discovery that one has an internal cancer. Much as the person suffering from coronary artery disease may be aware of the thumping of his scarred heart in the quiet of the night, he does not encounter the same demons that arise from knowing that one's body has given birth to a malignancy.

From time to time, the United States government has taken a few faltering steps to ease the plight of the million or so Americans newly stricken by cancer each year. During Lyndon Johnson's presidency, this took the form of regional medical centers for the treatment of heart disease, stroke, and cancer—whether a good idea that was never given a chance or a bad one, we'll never know, since it ended up as yet another casualty of the Vietnam War. The Richard Nixon administration actually moved the Congress to declare war on cancer, and many of the battles begun under this banner are still being fought. Under President Ronald Reagan, himself a colon-cancer victim, a much-needed reorganization of federally funded treatment centers has occurred but few new initiatives taken, although the billion dollars allocated to AIDS research may, by providing more information about tumor-producing viruses, speed progress in the cancer field, too.

On the surface, the seventeen-year-old "war" appears stalemated. Some statisticians, pointing to the increase in the cancer death rate from 170.2 per 100,000 Americans in 1962 to 185.1 per 100,000 in 1982, would even go so far as to call it a losing effort. Beneath the surface, however, in the research labs, if not yet in the treatment clinics, a strong sense of optimism prevails. As one eminent cancer researcher expressed it, "I've been looking down the barrel of a microscope at a cancer cell for the last thirty years and I think I just saw it blink!"

Until recently medical oncologists had to rely mainly on chemotherapy, whose impact Dr. Vincent DeVita, a pioneer in the field, aptly sums up by stating that "fifty percent of all cures through chemotherapy occur in ten percent of all cancer patients—mostly those suffering from the lymphomas and leukemias." But now, with a much clearer understanding of the enemy's ways and two spectacular new weapons technologies, monoclonal antibodies and recombinant DNA, the course of the war may have shifted sufficiently for one to wonder whether the time has finally arrived to make an all-out effort to win it. Put another way, if our government were to forgo the building of one nuclear-powered aircraft carrier and transfer the $1.5 billion budgeted for this purpose to the National Cancer Institute, how much of a difference would it make? It's beyond my knowledge to predict accurately, but I'd guess that in one area alone—that of monoclonal antibody therapy—it would accelerate the slow, steady progress now being made by at least five years.

Here's what makes me think so. As my Hurley Medical Center team prepares to enter a new treatment phase to test the effectiveness of much larger doses of monoclonal antibodies than previously used, we are sorely in need of answers to a dozen questions; without them, we will not know for years whether we did the best we could for each of our patients with what was available. To illustrate the scope of our dilemma, I offer, without too much in the way of technical explanation, the following short list:

- Are higher doses of the 17-1-A antibody—8–12 grams instead of 200–400 milligrams—sufficiently more effective against gastrointestinal cancer than the lower doses to warrant the increased risks?
- Does the procedure of leukopheresis, in which a portion of the patient's white blood cells are collected by a special machine and incubated with monoclonal antibodies before being infused back into the patient, contribute enough to the success of the treatment to be made a routine? Or can it be omitted, thereby enabling physicians to administer the antibodies in a clinic, or even in their offices, instead of in a hospital?
- Will "cocktails" of multiple monoclonal antibodies, each targeting a different site on the cancer cell, prove more effective than single-antibody infusions?
- Since, by some estimates, it takes four immune system killer cells ganging up on one cancer cell to ensure its destruction, should attempts be made to boost a patient's immune forces through pretreatment with one of the interleukins, interferons, or macrophage-stimulating factors before monoclonal antibodies are given?
- Must the "guided missile" type of weapons systems that these antibodies represent contain a "bomb" in the form of a radioactive isotope or a cellular poison, such as ricin, to be truly effective—and if so, which one?
- Should repeated infusions of monoclonal antibodies prove nec-

essary to control or cure most cancers, will "chimeric" antibodies—that is, half-mouse, half-human laboratory creations—solve the mouse protein allergy problem that somewhat restricts their use in this manner now?

- Should the antibodies be administered not only intravenously but directly into any body cavity known to contain cancer—as was done successfully with Jim Federline in France and a patient of mine with lung cancer more recently?

These are but a few of the pressing problems that must be solved before the best way to administer monoclonal antibodies and the true value of this form of immunotherapy can be determined. How long will it take? Most experts estimate ten to twenty more years. But even if monoclonal antibody technology fails to produce the decisive weapon against advanced cancer that those of us testing its products hope, in one area alone—that of diagnosing certain cancers at an early stage through blood tests and isotope scanning procedures—it has already saved many lives. An example is the 19-9 monoclonal antibody blood test that Wistar scientists have developed for the detection of gastrointestinal malignancies, especially the "silent" types like pancreatic cancer. Extensive clinical studies by Japanese investigators at the University of Osaka have revealed that the 19-9 antibody test is 78 percent accurate in diagnosing cancers of the pancreas at a stage when surgical removal and potential cure is still possible. Monoclonal antibodies to detect the presence of breast, ovarian, prostate, and other common cancers, and to signal their recurrence following treatment, have likewise been developed by Wistar Institute and other immunologists and are currently being evaluated. It shouldn't be long before physicians have available simple, sensitive tests with which to screen large groups of patients for malignant and even premalignant growths—and this advance alone will doubtless save innumerable lives.

Moreover, with Dorothee Herlyn and others at the Institute, Hilary Koprowski is working on a novel approach to the immunotherapy of cancer based on studies of surgical specimens from a large number of cancer victims. These show, in many instances, that the patient's immune system has attempted to mount an antibody attack against the malignancy in its midst but that it is woefully inadequate. To bolster it, Wistar scientists have developed a special vaccine derived from the patient's own antibodies that, in preliminary human trials, has proved capable of significantly boosting anti-cancer defenses. This is but one example of the new strategies that molecular biologists, through their creation of such awesome biotechnologies as monoclonal antibodies and recombinant DNA, have made possible in recent years, and why I remain optimistic that within another decade *most* cancers will be curable.

In the interval, though, what are cancer patients who cannot wait even a few more months for effective treatment to do?

Just about everybody has heard the saying "Money can't buy you love," and possibly its corollary: "Maybe not, but it sure helps you look for it in better places!" The same holds true for health; rich people, in general, get better medical care than poor people. Yet, as mentioned in the previous chapter, there are exceptions, and the cancer patient who has already undergone one experimental therapy and needs another is one of them. Or at least he was. As with almost everything else in a free-enterprise system, when a market demand exists for which people will pay generously, someone will usually find a way to provide it—and in Jim

Federline's case that someone appeared to be Dr. Robert Oldham.

There are two versions of why, after many years of valuable service to the National Cancer Institute, Dr. Oldham left to found his own treatment center. Oldham says it was mainly because of his dissatisfaction with the bureaucracy and delays and some of the selective interests of the Institute. Or as one breast-cancer patient put it, "If you've got cancer and need the help of NCI physicians to overcome it, you'd better have the type of cancer they're studying at the moment." What Oldham took issue with most, and set out to provide an alternative for, was, in his words, the "regulated monopoly" that the National Cancer Institute and the fifty-two "comprehensive treatment centers" affiliated with it exert over experimental cancer therapies nationwide.

NCI director Vincent DeVita sees the conflict differently. Says he of Oldham's decision to leave his employ and strike out on his own, "Dr. Oldham was of one mind and the rest of the world was not. He was in an incompatible atmosphere and so had to leave."

One of the more troubling issues in the dispute between Dr. Oldham and what he refers to as the "cancer establishment" involves patient access to state-of-the-art treatment programs. The usual way a cancer patient gains entry into a NCI-sponsored therapeutic trial is through referral by his private physician to the director of the study. Except for the costs of hospitalization, which may be borne by his insurance carrier, he is then treated free of charge. In return, he assumes the risks of a form of treatment that is either untested or unproven and whose side effects can be severe. He may even have to take his chances with a randomized study design, a coin flip by a computer determining whether he will receive the new therapy or an older, possibly less effective one. Moreover, this type of study usually tests a single treatment modality, such as Tumor Necrosis Factor or monoclonal antibodies, instead of a combination of agents that may prove more effective. For these reasons, the patient is not expected to bear any of the costs, which in the case of interleukin-2 plus LAK cells may run as high as $100,000 per treatment course.

The Oldham system gives the patient considerably more choice. The patient, if knowledgeable, can select the single or combined form of therapy that appears to offer the best chance of a cure, providing he or his insurance carrier is willing to pay for it. Thus, in a sense, both systems are exclusionary—the one because a patient may be unlucky in the type of cancer he has contracted or the remedy he has previously received for it; the other because of his inability to pay.

Dr. Oldham argues that, despite the cost factor, his institute fulfills a need by offering experimental treatments to a segment of the cancer population that might otherwise be denied them because of test subject restrictions. His critics counter that not only is it unfair and unethical to expect patients to underwrite the expense of experimental therapies but by so doing Dr. Oldham is "selling" research that others have developed. One critic asserts, "You can't buy high-quality research. You get high-quality research by going to high-quality research institutions," and in the main he is right—providing such institutions will take you in. Another concern of those opposed to patient-funded research is that it may not always be subjected to the rigorous peer review by federal and hospital authorities that help to ensure its safety and effectiveness. Theoretically, this is so, though I find a bit alarmist the warning of one critic that under such a system "it is conceivable that a patient's personal physician would be the only person looking out for that patient's interest." Has not a patient's personal physician always served as the main medical decision maker?

Although one might gather from the way I've slanted the foregoing arguments that I'm basically in favor of patient-funded research, I've also encountered its dark side—not, as it turns out, in dealing with Oldham's institute, but with a West Coast medical center. In brief, a patient of mine was found to have a type of inoperable lung cancer for which this medical center had developed a monoclonal antibody linked to a powerful chemotherapeutic agent. Unfortunately, at the time I contacted them, the researchers in charge of testing this "armed" antibody said that their supply was exhausted and they were awaiting approval of a government grant application, usually a lengthy procedure, in order to purchase the $40,000 piece of equipment needed to manufacture more. Were my patient able to donate this sum to their program, they would undertake to treat him—not an unreasonable condition, since I, not they, broached the subject of a donation, but one which would cost the man and his wife much of their life's savings. Having already been advised by several oncologists that the response of his type of cancer to chemotherapy was so poor that they no longer used it, my patient was in a quandary: was there any alternative to his having to raise the $40,000?

Happily, there was. My research partner, Harland Verrill, tested a specimen of his tumor against the panel of Wistar Institute-produced antibodies we had on hand and found one that reacted strongly against it. Zenon Steplewski generously furnished us with his entire supply of the antibody, and in September 1986 the patient received it. Comprehensive X-ray studies four, eight, and fourteen months following treatment showed no trace of his previous disease, and as of this moment he remains cancer-free.

Each side in the growing controversy over fee-for-service research centers its argument on the patient. Certainly for a cancer victim to be denied a chance of a cure because he does not fit a drug-testing protocol is a crushing blow. Yet, in the view of one bioethicist, it might be morally preferable for the therapies offered by Dr. Oldham's institute to fail. "Should they succeed," writes Dr. Thomas Murray of the University of Texas, "we will have a situation where . . . the wealthy live and the poor die." The same, however, might have been said about kidney transplants at one time and heart and liver transplants more recently.

Since, to date, Dr. Oldham's institute has treated less than a thousand of the hundreds of thousands of victims of advanced cancer, one might wonder what the fuss is all about. The answer, I suspect, has less to do with ethics or science than *control*—control of experimental therapies for the individual patient and for cancer sufferers as a whole. This is pointed up by the fact that almost a third of Dr. Oldham's patients are either physicians themselves or members of their immediate family. This group, perhaps better than most, can weigh the possible risks and benefits of the latest cancer remedies and are willing to pay a substantial sum in order to have some say in their treatment. Much as National Cancer Institute officials, as well as thoughtful physicians everywhere, might justifiably worry whether the entry of private companies into the cancer field will make as big a mess of it as they now appear to be doing to our hospital system, this consideration is understandably of little concern to the physician-patient struggling to overcome cancer. Yet, as one of them admitted, "I feel a little guilty having a chance at recovery that others could not afford."

Whether Jim Federline had similar thoughts while consulting with Dr. Oldham in late July, I never asked; but I do know that, despite the discouraging outcome of the visit, he was favorably impressed by the skill and sincerity of Oldham and his associates.

In a phone conversation with Jim and later in a formal consultation letter from Dr. Oldham, I learned that, because of the large amount of fluid in Jim's abdomen and the propensity of interleukin-2 infusions to worsen this complication, he was not a candidate for this form of therapy. "Nonetheless," Dr. Oldham wrote, "I appreciated the opportunity to have seen Mr. Federline. It was an education for me to see how a single individual can gain access to so many experimental programs and work so hard on his own behalf."

Having correctly anticipated what Oldham's decision regarding interleukin-2 therapy would be, Zenon Steplewski had told Jim that there was one more treatment method worth considering, though it would be doubly dangerous: not only would a fourth infusion of monoclonal antibodies carry the likelihood of an allergic reaction, but the radioactive iodine isotopes that Zenon planned to attach to them would subject Jim to a heavy dose of radiation. With Jim reporting his condition as stable for the moment, he was advised to wait a while longer before deciding whether to risk this form of treatment, especially since Zenon's team could use the extra time to find the best antibody with which to deliver this lethal bombardment to his tumor.

17

JIM FEDERLINE was out on the deck of his Gaithersburg, Maryland, home, a telephone within reach and a company report in his lap, when the strange sensation struck. Suddenly his jaw began to tremble and he experienced the same sort of abrupt, aching skin changes as if he had walked directly from a sauna into a cold-storage room. An early-afternoon sun blazed overhead and the temperature on this end-of-July day was well into the nineties, yet Jim's teeth chattered, his limbs shook, and he was racked by shivers. Struggling up from his chair, he moved into the sun, but still couldn't seem to get warm. His legs felt so rubbery that he had to hold tight to the deck railing. Invisible pins pricked his skin. The shaking of his head and limbs grew so violent that he felt as if he were atop a mechanical bull on open throttle. Was he having some sort of epileptic seizure? he wondered. Had his cancer spread to his brain?

Tinka was nearby and responded at once to his call for help. Rapidly, she put Jim to bed and covered him with extra blankets, took his temperature, and phoned Dr. Fred Smith.

A chilly sensation commonly heralds the onset of a mild viral infection. But a truly incapacitating, teeth-chattering chill almost always means bloodstream invasion by disease-producing bacteria, and this is what Dr. Smith found—the likely source being the external end of the plastic catheter in Jim's neck vein through which he received the intravenous nourishment to help him maintain his weight. Much as Jim disliked hospital confinement, he felt too weak to offer more than a token protest when Dr. Smith insisted on it.

Jim spent the next ten days in Washington's Sibley Hospital to complete a course of high-dose, intravenous antibiotics. While there, he also underwent a series of X-ray studies that revealed his tumor mass to be approximately the same as a month before, having neither shrunk in size nor spread to his liver or lungs. Jim did not need to view the CAT scan to know that the fist-sized lump in his abdomen was still there, but he liked to think that most of it was dead.

Jim felt much improved on leaving the hospital and was determined, by whatever means possible, to regain his strength. Tinka, fearful that he might become infected again, redoubled the antiseptic precautions she took with his intravenous feedings. More and more the room in their house set aside for this purpose grew to resemble a hospital nursing station. Each morning Tinka weighed Jim and wound a tape measure around his abdomen; three times a day she took his temperature and with meticulous care connected a bottle of nutrients to the Portacath in his neck. To keep their two young children from becoming even more frightened and worried about their father than they already were, Tinka invited them to watch whenever she changed the bottles and Jim's bandages and made something of a game of it by giving each piece of equipment a nickname, such as "Andy" for the infusion pump.

Jim's first week home from the hospital went well, and the

Federlines spent much of the second preparing for the big party and family reunion celebrating Jim's mother's seventieth birthday that they would give the following Sunday. Three days before, however, Jim woke feeling unaccountably weak—the reason becoming clear shortly when he suffered a shaking chill. Immediately Tinka bundled him up and drove him to the emergency room of Sibley Hospital, where Dr. Smith strongly recommended that he be admitted, have the site of his Portacath changed, and receive another course of antibiotics. But Jim, wanting to be home for his mother's birthday celebration, refused, and a compromise was worked out: for the next three days Tinka would administer the antibiotic Dr. Smith prescribed at frequent intervals through Jim's Portacath and, provided his condition did not worsen, he would put off entering the hospital until after the party. Though Tinka shared Dr. Smith's uneasiness about this arrangement, she knew Jim was motivated by his belief that seeing him at home, instead of in the hospital, would reassure his mother that his battle with cancer could yet be won.

Jim did attend the party, although it proved to be an ordeal. Feeling extremely weak and, at times, faint, he could mingle with his guests for only brief periods, and every few hours Tinka excused herself to go to the bedroom and administer his antibiotic. Each time she did, her apprehension over Jim's condition worsened. The moment the last guest left, she drove him to the hospital.

Jim spent the next week on the oncology floor of Sibley Hospital, again responding well, if more slowly, to antibiotics. This latest setback acted to dispel whatever misgivings he harbored about undergoing the high-risk treatment Zenon Steplewski had suggested earlier. While still in the hospital, he phoned to ask my advice about it, and on the day he was discharged he called Zenon to make the arrangements.

To fell a charging elephant with a small-caliber rifle requires that the ill-equipped hunter have intimate knowledge of the animal's anatomy and squeeze off an accurate shot. In the same way, for a monoclonal antibody to cripple a cancer cell and make it easy prey for the immune system, it must attack the cell at a critical and vulnerable spot.

What makes the Wistar Institute's 17-1-A antibody effective against many gastrointestinal cancers is its ability to home in on a protein of uncertain function present in much greater density on the surface of these malignant cells than on their normal counterparts. Though the same protein exists in normal intestinal cells, it is at a subsurface location, too deep to attract the antibody.

By 1986, it had become apparent to cancer researchers that the principal action of many oncogenes involves the production of excessive numbers of various growth factor receptors. It was equally clear to Wistar Institute scientists that one way to make monoclonal antibodies better cancer killers was to program them to attack these membrane receptors. A precedent for this approach existed in the increasingly successful efforts of National Cancer Institute immunologists to fashion a monoclonal antibody treatment for victims of T-cell leukemia. Earlier work by Dr. Robert Gallo's group had shown that a major difference between normal and malignant T-cells lay in the number of interleukin-2 (T-cell growth factor) receptors on their surfaces: 10,000 or more for leukemic cells versus fewer than 500, if any at all, for normal cells. On the basis of this finding, the NCI team headed by Dr. Thomas Waldman created a monoclonal antibody against the interleukin-2 receptor, administered it to seven leukemia victims, and obtained significant falls in the leukemic cell counts in two. To make these antibodies even more potent, Dr. Waldman has linked them to various radioactive isotopes and cell poisons and is currently testing them in the clinic. But even if this strategy proves highly successful, the thirty-year-old chemotherapy experience has shown that what works for blood-cell malignancies may or may not work for solid cancers under a different hierarchy of cell-cycle controls. Nonetheless, for at least one member of the solid group, malignant melanoma, the findings of Dr. Meenhard Herlyn and his Wistar Institute co-investigators strongly suggest that it might.

Of the half million Americans who develop skin cancers each year, only around 8,000 die of them—7,500 from melanomas. Not only is this the deadliest form of skin cancer but its occurrence is sharply on the rise. Melanomas stem from the pigment-producing layer of cells called melanocytes, which protect the body from harmful ultraviolet rays, determine hair and skin color, and incidentally cause much racial prejudice. These cells give rise to a variety of pigmented tumors—unfortunately, including melanomas—that are ideal for study since they are visible to the unaided eye and accessible to serial skin biopsies. From such investigations, Dr. Herlyn has discovered that, as a clone of melanocytes progress from benign moles to melanomas, they pass through five stages, the most critical of which correlates with a steep increase in the number of membrane receptors for various growth factors. Of particular interest is the one known as epidermal growth factor, since one of its main sites of action is the epidermis of the skin. Unlike full-blown melanomas, neither harmless skin moles nor early-stage melanomas, which do not possess the wherewithal to spread much beyond their birthplace, display many epidermal growth factor receptors on their surfaces. In short, the more malignant the melanoma, the more its oncogenes express receptors for the external stimuli needed to maintain its abnormal rate of growth. Like the elephant hunter, cancer researchers hope to use these receptors as the bull's-eye to stop a melanoma dead in its tracks.

From a cancer cell line known to overproduce a portion of the epidermal growth factor receptor complex, Dr. Herlyn's group has created a monoclonal antibody that attacks not only melanoma cells growing in culture dishes or implanted in nude mice but certain breast, colon, and brain cancers as well. Having heard about this from Hilary Koprowski, I paid a visit to Meenhard's lab the third Monday in September, the same day that Jim Federline arrived at Philadelphia's Hahnemann Hospital for further treatment. With Jim in mind, I asked Meenhard if his anti-receptor antibodies were ready for patient trials; they were not. Neither, as it turned out, were the antibodies loaded with radioactive iodine with which Zenon Steplewski, through the Hahnemann Hospital radiology group, had hoped to treat Jim Federline. In lieu of this—and because something obviously had to be done to halt Jim's rapid deterioration—it was decided to try a combination of therapies whose rationale was different.

In order to achieve maximal cancer-cell kill, monoclonal antibodies not only have to unerringly seek out their target but must pass intact through the endothelial lining of the capillaries feeding the cancer and whatever barriers it has erected against the immune system. Since preliminary results from animal experiments indicated that X-ray bombardment increases the numbers of antibodies reaching a cancer, the combination of the two treatment methods seemed worth a try—though the gains, if any, would exact a heavy price. As was explained to Jim, the radiation treatments would likely produce intense nausea; this, plus the almost certain recurrence of the severe pain he experienced the last time monoclonal antibodies were infused directly into his abdomen, meant he would be in for a very rough time. No cure was even contemplated; but if his "luck" held and he suffered neither an allergic reaction to the mouse monoclonals nor a bowel perforation from a cancerous patch on his intestines burnt out by the X-ray beam, it might buy him a little extra time.

Even when tempered by staunch optimism, the proposed treatment sounded like pure misery—attractive only in the light of its morbid alternative. Yet with Tinka's support, Jim managed to view it as one more step down a never before traveled road that might still lead to a surprising and desirable place.

Before flying home to Michigan at the end of that week, I phoned the Federlines at their Philadelphia hotel and spoke to Tinka, since Jim, who had just returned from Hahnemann Hospital and his sixth X-ray treatment, was napping.

With his usual fortitude, and an occasional painkiller or anti-nausea pill, Jim had endured his first week of radiation treatments with few complaints. Yet Tinka was more than a little concerned because his abdomen was rapidly enlarging with fluid, adding almost an inch a day to his girth. And for the past two mornings Jim had been wakened by a fit of coughing, producing a brownish phlegm, that left him breathless for several minutes. Though the therapeutic plan called for a total of ten X-ray treatments before the infusion of monoclonal antibodies into his abdomen, Tinka was uncertain whether Jim would be able to meet this schedule. The store of energy with which he had begun the week was depleted and he was feeling extremely weak. She planned to take him home for the weekend; then, if he felt strong enough, he would return to Hahnemann to resume the radiation treatments.

From Tinka's description of his debilitated state, I figured it would take Jim much longer than a weekend to regain his strength. Yet, the following Monday, the Federlines drove back to Philadelphia, checked into the same hotel, and with the sole exception of taking a taxi instead of walking the six blocks to the hospital, carried on as before.

On Thursday of the following week, I phoned Jim at Hahnemann Hospital to find out how his monoclonal antibody treatment had gone. Since he was heavily sedated at the time, Tinka took the call and, though she spoke calmly, what she said gave me great cause for concern.

The previous day, upon receiving an intraperitoneal infusion of the 17-1-A antibody, Jim had experienced some pain but no other difficulties until later that evening when he suffered either a delayed allergic reaction or some sort of cancer-related heart complication—his doctors remained uncertain which—and began gasping for breath. Tinka rushed him back to the hospital, his lungs so full of fluid that two quarts of it had to be drained by needle. Since then, though his abdomen felt very tight and sore, Jim's breathing difficulty had largely abated, and despite the setback, Tinka assured me that he had lost none of his fighting spirit. To bolster it further, she hoped he would be allowed to travel home for at least a day on the weekend to visit with the children.

18

IF ASKED to name the most troublesome issue I face, day by day, as a physician, I would immediately answer, "Everybody wants to go to heaven, but nobody wants to die first!" The words are not mine but those of the late heavyweight champion Joe Louis, and they best express the seductiveness of death. There comes a time in the care of many critically ill patients when, for practical or humane reasons, they ought to be given the opportunity to go on to what one hopes will be a happier existence without further treatments or delay.

Had that time come for Jim Federline? I wondered. A year earlier I might have thought so. But since then, the responses of several cancer patients to monoclonal antibody therapy—two in particular—have encouraged me to keep trying.

In mid-November 1985, shortly before I first met Jim Federline, Dr. Rao Mukkamala, a talented radiologist on the Hurley Medical Center staff, asked me to treat a psychiatrist friend of his from Cleveland whose abdominal CAT scan revealed a tumor the size of an orange in the area of his pancreas. Despite this, Rao assured me, his friend continued to work and was eager to try our new treatment.

Dr. Conjeevaram Ramamurthy—"Rama" to friends and patients alike—entered Hurley Medical Center on November 17, 1985 and related to me what, in many respects, had now become a familiar story. Except for coronary heart disease that had responded well to coronary artery bypass surgery in 1982, Dr. Rama had enjoyed good health and a busy psychiatric practice until early 1985, when he began to grow progressively more fatigued and to lose weight rapidly: twenty pounds in a single month. In Rama's words: "My body was trying to tell my mind that it was in trouble, but in the absence of pain my mind refused to listen." Then came a chilling moment of self-discovery identical to that experienced by Jim Federline. In bed one night, Rama began palpating his abdomen and felt what he thought was an enlarged liver. The next morning, he visited his personal physician, who ordered a liver scan. To Rama's relief, this proved negative, but it left his symptoms unexplained. When his fatigue grew so profound that he had only enough energy to see his patients in the morning, Rama underwent further diagnostic tests, including an abdominal CAT scan, that revealed his tumor and led to exploratory surgery.

The news that he was suffering from an inoperable pancreatic cancer shocked Rama. Depression, a disease that, through his patients, he had come to know well over the years, descended on him, and for weeks after the operation he struggled against it. Finally he mustered the determination to undergo a five-week course of combination radiotherapy and chemotherapy with the drug 5-fluoruracil. As a side effect of the radiation treatments, his stomach became severely inflamed and Rama suffered protracted bouts of nausea and vomiting until the inflammation eventually subsided. Around this time, his friend Rao Mukkamala paid him a visit and told him about the monoclonal antibody trials we were conducting at our hospital. Laboratory tests showed that, when incubated together and examined under the microscope, the 17-1-A antibody reacted strongly against a specimen of his cancer.

In the months following his treatment with the antibody, Rama improved rapidly. His fatigue abated, he regained the weight he had lost, and he resumed his psychiatric practice part-time. When a repeat CAT scan two months later revealed no change in the size of his tumor, Rama was philosophical: he was feeling so well, physically and mentally, that even if the X-ray images did not reflect it, he knew he was winning the war against his cancer. What he found especially encouraging was that he could no longer feel a mass in the area of his liver.

Science, however, demands more objective measurements, so a third CAT scan was taken in October 1986. It revealed no significant change, and Rama has not had another such examination since. It is not that he is afraid to know what the test might show, Rama assured me in our last phone conversation; he simply is not convinced it is necessary. After all, he points, out, over two years have now passed since the monoclonal antibody treatment; his energy level is good, there are no lumps in his abdomen, and he continues to work—what better proof can there be of a favorable response?

Dr. Ramamurthy's response was one reason why I did not discourage Jim from seeking further treatment. The other stemmed from an experience so personal that I have been reluctant to write about it in its true context until now. What restrained me was superstition based on a fearful appreciation of irony—the same sort of superstition that still forbids baseball broadcasters from even hinting that a pitcher has a no-hit ballgame in progress until his team makes the final put-out. Nonetheless, enough time has passed since I first treated my closest friend for cancer that I now feel I can risk it.

I first met Kenneth Kay and his wife, Charlene, in 1957, when I was a U.S. Army medical officer assigned to Supreme Allied Headquarters in Paris and Ken was an Air Force colonel stationed at a large American military base nearby. In addition to his skill as a logistics officer, Ken was a highly regarded fiction writer who had just published his first novel, and I was a physician with similar aspirations. At the time, Ken had a few minor health problems and I had more than a few problems with the novel I was trying to write. He needed a physician; I needed a writing mentor; and it seemed only reasonable to help one another out.

Ken retired from the Air Force in 1966, settled in Tampa, Florida, and joined the faculty of the University of South Florida. In his capacity as adjunct professor of English, it was his duty to teach a generation of students, whose knowledge of the world came more from TV screens than from books, how to write properly. That he succeeded in this mission is attested to by the dozens of Professor Kay's former students who have gone on to writing careers; I consider myself one of them.

In 1980, Ken and I co-authored a novel, *Disposable People*, that posed a provocative question (all the more so since the outbreak of the AIDS epidemic): What if a modern black plague were to erupt in the United States for which no effective treatment existed; and what if the government needed a host of human subjects on which to test a risky experimental vaccine? Following the book's publication, Ken decided he wanted to write another medical thriller that would incorporate some of his experiences as an Air Force officer in China during World War II and asked for my help with the medical portions.

This was shortly after I'd begun working with Wistar scientists to evaluate the accuracy of the blood test they had developed for the diagnosis of gastrointestinal cancer, and monoclonal antibodies were very much on my mind. Hence, the plot of Ken's novel revolves around a Chinese premier stricken by cancer. The United States government feels it is to its advantage to try to keep the man alive, especially since his vice premier's sympathies are overtly pro-Russian. In secrecy, a specimen of the Chinese leader's cancer is flown to a U.S. research institute, monoclonal antibodies made against it, and the premier treated with them. The book is entitled *The Man Who Must Not Die.*

Ken completed the novel in the spring of 1986, and its publication was scheduled for the fall of 1987. "Then came the cruelest irony of my life," he writes in a preamble to the book.

In June 1986, Ken developed difficulty in breathing and was discovered to have a pleural effusion: a collection of fluid filling the lower half of his left lung. At first, his physician thought—and I hoped—that it had been caused by a drug reaction. Ken's first hint that it might be something more serious came from a pulmonary specialist who advised him that, even though Ken had quit smoking tobacco fifteen years before, the likelihood was that he had lung cancer. Extensive tests to confirm this diagnosis were inconclusive, until an open-lung biopsy showed cancer nodules studding the inner lining of Ken's chest cavity on the left side. Had the cancer been confined to the involved lung, operative removal might have been possible. But since it had already spread to the inner chest wall, a surgical cure was ruled out.

Making no effort to hide his despair, Ken phoned to tell me the bad news and ask my advice on what to do next. Though my experience with monoclonal antibody therapy was limited to gastrointestinal cancer patients, I promised to find out what I could about their use in carcinomas of the lungs. The following day, Ken telephoned again. A highly reputable oncologist, Dr. Rand Altemos, had been to see him and candidly admitted that the results with chemotherapy were so poor with his type of lung cancer that the oncologist no longer recommended it. Much as the prospect of undergoing chemotherapy had depressed Ken, the thought of no treatment at all was worse; without it, he could muster scant hope.

From Zenon Steplewski, I learned of a California cancer clinic conducting clinical trials with a monoclonal antibody against Ken's type of lung adenocarcinoma. But the clinic's supply was exhausted, and even if Ken helped to pay the production costs, it would still take them a minimum of four months to make more. That moment was a low point for me, but instead of telling Ken this, I told him that Harland Verrill and I had decided to become more actively involved in his case. We obtained a specimen of his cancer, tested it against the Wistar Institute monoclonal antibodies we had on hand, and discovered, to our delight, that one of them, the 55-2 antibody—a monoclonal that had been administered to only a few dozen patients before—reacted well against his cancer. Zenon Steplewski volunteered his entire supply, and the following week Ken flew to Flint to receive it.

My wife, Barbara, and I had last seen Ken the summer before, at which time he looked his usual robust self. But when Barbara went to meet him on his arrival at the Flint airport she was shocked by the changes in him. Here was a man who, up until a few months ago, had bicycled many miles a day, but whose lung power was now so impaired that he required a wheelchair to transport him the hundred feet from the airline gate to my wife's car. When I came to see him in the hospital, I was equally upset by his wasted appearance and made no attempt to hide it. "You look like hell," I said as we hugged, knowing Ken would cringe at any offering of false cheer. "I feel like hell," he replied. "What can you do about it?"

I outlined the treatment schedule we planned to follow: leukopheresis and monoclonal antibody infusion the first day; an injection of antibodies directly into his left lung the second day; and a repeat intravenous infusion the third day. Such intensive therapy with an experimental antibody that had never before been used to treat lung cancer carried a risk, I warned, but Ken shrugged it off: it was the best offer anybody had made him thus far.

Ken tolerated the three-day course of treatment well and at its completion returned to Tampa to await results. I had cautioned him not to expect any significant improvement for at least a month. Two weeks later, however, he phoned to tell me that he was feeling considerably more energetic and had regained some of his lost weight. His phone call the next week brought startling news: Ken had begun riding his bicycle again, a mile or so an outing, and was confident he could improve on that distance. Within a month, he was bicycling ten miles a day, an impressive performance for most seventy-two-year-old men, but especially so for one who, four weeks earlier, could not walk a hundred feet.

In January 1987, Ken returned to Flint for follow-up X-ray studies and, if necessary, further treatment. I awaited the completion of his CAT scan with great trepidation, knowing as I did that an additional supply of the 55-2 antibody would not be easy to obtain. With Ken waiting tensely outside the radiology office, I met with Rao Mukkamala to review his scan. "It's negative,"

Rao said the moment I entered the office. "What do you mean, negative?" I asked excitedly. "There's a lot of scarring in his left lung, but very little fluid and no sign of cancer," Rao replied.

Immediately I invited Ken in to look at the serial cross sections of his lungs as they appeared on the computer console's screen. Ken was both fascinated and jubilant. On leaving the radiology office, a thought struck me that neither of us had had before. "Good Lord," I exclaimed to Ken, "do you realize what's happening? You're actually living the plot of *The Man Who Must Not Die*!"

As Ken writes in "a personal note to the reader" at the beginning of his recently published novel:

> That was seven months ago [September 1986]. Within days of returning home, I mounted my bicycle and pedaled off, doing nine-tenths of a mile the first day. I built up gradually and now ride ten miles almost every morning. My appetite surged back. I regained twenty lost pounds and am beginning to have to watch my diet; most dramatic of all is that within a month my diseased left lung, squeezed to uselessness by tumor-induced serous fluid, began breathing again in a limited way. After three months I flew back to Flint—no wheelchair this time!—where X-rays, CAT scan, and other tests showed no sign of tumor or serous fluid and that the diseased lung had expanded to about a third its normal size.
>
> *The Man Who Must Not Die* is pure fiction. Its characters and events are all imaginary. But as my own experience has shown, the revolutionary medical treatment it tries to depict is far from imaginary and is growing less so every day.

19

ON SEPTEMBER 30, 1986, five days after my last phone conversation with Tinka Federline, I returned to my office from making morning hospital rounds and had barely sat down when my secretary, Wilma, looking unaccountably upset and pale, entered, stopped well short of my desk, and stared sadly at me.

"Jim Federline died late last night," she said. "His secretary, Sally, phoned a few minutes ago to make sure you knew."

The surprise announcement of the death of someone you know well stops the breath. I was no different. "Where?" I asked.

"Aboard an airplane on the runway of Washington's National Airport."

"An airport?" I repeated, my mind conjuring up a scene of mist-enshrouded landing lights, the whine and whoosh of jet engines, the piercing wail and flashing lights of an ambulance cutting across the tarmac. An eerie place to die; a death probably as dramatic as the last thirteen months of Jim's life.

Softly Wilma asked, "Would you like me to try to reach Tinka Federline?" and at my nod she left.

While waiting for the call to go through, I managed to clear my brain enough for some searching self-examination. Surely,

Jim's death came as no great surprise, I told myself. A shock, yes, but no surprise. As an experienced physician, I must have known the Thursday before that he was going to die, *and I did know*; my problem was that I refused to believe it. Jim had survived so many previous crises that I'd just assumed he would survive the latest. But now that this illusion had been demolished, replaced by thoughts of the terrible pain and torment he had suffered for so long, I felt compelled to ask myself if anything truly worthwhile had been accomplished by all our experimental treatments.

Except for the pursuit of new knowledge, I had no good answer. Fortunately, at the end of the brief, sad conversation I had with Tinka, she gave me one. "Jim and I needed that extra year," she said. "We became very close."

Sigmund Freud wrote, "It is impossible for an individual to accurately envision his own death; for, however hard he tries, he still exists as an observer." My own occasional morbid musings tell me that Freud was right. Yet the difficulty I was having now was not in contemplating my own death but Jim Federline's. From Tinka, I learned that Jim had expired around ten the previous evening on the private ambulance airplane she had hired to bring him home to die. But beyond this, I lacked the details that might help to impress upon me the reality of Jim's death.

It was not until months afterward, when Tinka felt ready to talk about it—first in a lengthy phone conversation, later in a tape recording she sent me—that I learned of Jim's last days, and hours, and finally came to accept the end of a brave man, a friend, and an experimental subject whose case number and therapeutic result would subsequently appear in a medical report under the heading of "unusual stability of advanced disease" ending in death.

Much of what Tinka said and wrote about the circumstances surrounding Jim's death was personal. But with her permission and by way of summing up, I quote the following passage from her February 27, 1987, letter to me:

> Jim knew that the odds were against him but in fighting as hard as he did, it gave us more time together than was first anticipated. During this thirteen months, we were able to develop a relationship closer than most people ever have in fifty years of marriage. We had time to talk, listen, and care for one another. He gave me strength that I didn't know I had to deal with a situation that otherwise would have been unbearable.

When Jim died, I had barely begun this book and so had a choice. With little in the way of lost effort, I could abridge Jim's case history and focus more on such monoclonal antibody success stories as that of my friend Ken Kay, who has been free of any sign of lung cancer for fifteen months now, or that of Norman Arnold, who has passed beyond the "five-year cure" mark for pancreatic cancer with ease. But the truth is, I never even considered such a change. One reason was that Jim's story is far more representative of the current reality: though time can often be gained for the victims of advanced solid cancers, much of it enriched by love and self-discovery, only a small proportion can be cured. Though I remain confident that medical science will soon find ways to reverse this situation, as of now the failures far outnumber the triumphs achieved by monoclonal antibodies or any other means of therapy.

That was one factor in my decision to adhere to my original intention to tell Jim Federline's story in full. Another was this: Years earlier, Jim and three of his more adventurous friends—one an experienced mountain climber—joined together to custom-order a sixty-two-foot oceangoing sailboat. Whimsically, and ironically, they christened it *Saga*, short for Sagarmatha, the Nepalese name for Mount Everest, not the English word denoting "a lengthy tale of heroic exploits"—though they hoped this definition might someday apply, too. Except in spirit, Jim can no longer be at the helm of the craft he helped to design. Yet the boat sails on—and in a way so does the Federline saga in the person of Tinka, a recent appointee to the Wistar Institute's board of managers, and the many cancer patients who have since benefited from the lessons learned in treating Jim. For the story that began for me on November 15, 1985, there is no end in sight.

Epilogue

IN MY LAST visit to The Wistar Institute before winding up the research on this book, I went looking for Hilary Koprowski and, acting on a tip from his secretary, found him in the Merieux Room—named for the French pharmaceutical company that helped to furnish it. He was seated at the grand piano kept there under lock and key for in-house concerts. At his side stood Marian Filar, his 1936 Warsaw Conservatory classmate and Temple University music professor, who was helping Hilary "practice his fingering." One intrudes upon such a private session with trepidation, but after hearing me say, "I'm not here to talk, just to listen," Hilary smiled his consent and kept on playing.

Hilary's favorite composer is Frédéric Chopin—not merely because of their common Polish heritage but because Chopin's musical compositions are so intricate and technically advanced. As he practiced a Chopin piece in stops and starts, I glanced around the room I had been in many times before but had never really examined. On one wall hung a huge tapestry, "The Triumph of Alexander the Great," and opposite it a life-sized mural of Hilary and friends lounging on a lawn. Much of the floor was taken up by a long conference table seating thirty, a blackboard, a slide projector on a stand, and a viewing screen. An intriguing contrast of Art and Science, I mused, and was reminded of British writer Anthony Powell's masterwork, *A Dance to the Music of Time,* which chronicled in exhausting detail an entire generation of Englishmen. But it was not so much the subject matter as the title of Powell's multi-volume work that my thoughts fixed on now. What, I wondered, if such a translation were possible, would the Music of Science sound like?

The next day, I posed the question to Hilary. Might science, I ventured, take the same form, movements, and tempo as the *Trois Nouvelles Etudes* by Chopin I'd heard him practicing the previous afternoon?

Hilary pondered and nodded.

"Describe it for me," I asked, and after brief reflection, he did:

"Slow and deliberate at the start. Romantic. Inventive. Aggressive. Stormy. Then reevaluation, reexamination. Conflict. Upsets . . . It is, after all, based on a love story."

Glossary

by Harland Verrill, Ph.D., Hurley Medical Center

antibody A Y-shaped protein molecule produced by a type of white blood cell, the B-lymphocyte, to react specifically against a given antigen.

antigen Any substance (virus, bacterium, etc.) that the body's immune system perceives as foreign and manufactures an antibody to react against.

arteriosclerosis A disease process which results in the buildup of fats, principally cholesterol, in the arteries of the body, restricting the flow of blood through these vessels.

autocrine This term is based on the term "endocrine," from "endocrinology" (study of hormones). Hormones are made by glands distant from the target organs. In an autocrine situation, the tissue itself makes hormones or growth factors for which it possesses its own receptor sites. Hence, it is self-stimulating.

biopsy The removal and examination of tissue or cells from the living body.

b-lymphocyte A white blood cell that is capable of making antibodies.

catheter A flexible tube introduced into a blood vessel or body cavity for the purpose of infusing or removing fluids.

cell division The process by which all organisms increase their numbers for growth and replacement.

cellular immortality Unlike normal human cells, which can replicate only approximately fifty times in a properly maintained culture dish, cancer cells, for unknown reasons, can reproduce themselves endlessly under such circumstances.

chromosome An array of many thousands of genes that store and transmit genetic information. There are 23 pairs of human chromosomes, identical copies of which exist in every cell of the body.

cytotoxic T-cells The subset of T-cells responsible for the killing of diseased or obsolete cells.

DNA The abbreviation for deoxyribonucleic acid. This double-stranded molecule is the material containing the genetic message.

endorphins A family of chemicals synthesized in the nervous system that act like morphine. These are often referred to as natural painkillers.

enzyme An enzyme is a protein that allows a chemical reaction to proceed rapidly in a biological system.

epinephrine (adrenaline) A hormone that is naturally secreted by the adrenal glands, which are located on top of the kidneys. This hormone can also be used medically to treat acute allergic reactions. It works in part by maintaining blood pressure and glucose production during stress.

gamma-interferon Gamma-interferon is produced by activated T-lymphocytes. It is also called immune interferon in that it mediates many immunological functions.

gastrointestinal Of or pertaining to the organs of food digestion. Usually the stomach, pancreas, and intestine.

gastrointestinal cancer Cancer predominantly of the colon, rectum, stomach, and pancreas.

gene transplants Procedures to provide a missing protein by inserting its missing gene into the defective cell. In humans, this is often done by bone marrow transplantation.

growth factor An external chemical stimulant for cell division.

growth-factor receptor In key-in-lock fashion, this protein structure on the surface of a cell serves as the lock for a growth factor or hormone.

hybridoma A hybrid cell created in the lab by the fusion of a cancer cell (which is "immortal" in the sense that it can reproduce endlessly in a test tube) and a lymphocyte previously immunized (or programmed) to attack a specific antigen. Hybridomas are literally "cellular factories," producing limitless amounts of monoclonal antibodies directed against a target on the surface or the interior of a cell. Once made, no one ever has to create that particular hybridoma again.

immunocytes Any white blood cell of the immune system that protects the body from harmful invaders or cancers.

immunoglobulin A protein that comprises the main structure of an antibody.

interferons A diverse group of small proteins made by immune system cells to protect healthy cells from viral infections and certain cancerous conditions.

interleukin-1 Macrophage-derived factor that promotes activation of the entire immune system, especially T-cells.

interleukin-2 Also known as T-cell growth factor. A chemical that promotes the proliferation of T-cells.

leukemias An acute or chronic white blood cell cancer characterized by an abnormally high number of white cells in the blood and tissues.

leukocyte The white blood cell. The body's primary defense against infection.

leukopheresis A procedure whereby blood is taken continuously from a patient and separated into its component parts. Red cells, white cells, platelets, and plasma are separated by the leukopheresis machine. The white cells are saved and the other components are returned to the patient.

lymphocyte A cell of the immune system which resides in the bone marrow, spleen, and lymph nodes. There are two types of lymphocytes; B-lymphocytes, which make antibodies, and T-lymphocytes, which control the output of antibodies by the B-cells and can destroy diseased cells directly.

macrophage A large immune system cell with the ability to ingest foreign invaders and kill diseased cells.

mitotic Refers to the appearance of condensed chromosomes when a cell is undergoing preparation for cell division.

monoclonal antibody Protein substances manufactured by hybridomas that are all exact duplicates (clones) of one another. Nothing in nature can copy so exactly.

Natural Killer cell An immune system cell that, along with macrophages, provides the body's first line of defense against viral invasion and cancer.

oncogene One of a family of fifty or so known cell-cycle-controlling genes which may lie dormant until activated by a virus or carcinogen that alters its on-off control regions.

oncologist A physician who specializes in the treatment of cancer.

pancreas The gland that lies on the posterior wall of the abdomen and secretes digestive enzymes and insulin.

paracrine The prefix "para" means beside or beyond. In describing systems for stimulating cells, a paracrine system is one in which one cell has the proper receptor and can respond to the stimulating chemicals produced by its neighboring cells.

phagocyte A cell that characteristically engulfs waste material and consumes the debris of dead organisms, such as bacteria and viruses, as well as dead cells.

radioactive isotope The form of a natural element that is electrochemically unstable and emits radioactivity. In the laboratory these isotopes can be used to measure very small amounts of biological substances. At high energy levels, radioactive isotopes can be used to visualize and in some cases to destroy cancer cells selectively.

RAS RAS is one of a family of oncogenes. These genes are usually not turned on and are dormant throughout life, except to assist rapid natural healing.

recombinant DNA technology The science of cutting and splicing together genes through the use of special enzymes and techniques.

recombinase An enzyme that can join genes, sets of genes, or parts of genes into new combinations.

retrovirus Viruses containing RNA instead of DNA as their basic genetic material. During viral reproduction, the genes of the virus become integrated into the victim's chromosomes. In this way the virus uses the host cell to replicate itself.

serotonin A nerve-impulse-transmitting chemical involved in many brain functions.

T-cell receptor T-cells are special lymphocytes which mediate cell defenses. The T-cell has receptors on its surface which match proteins bound to foreign cells. When these T-cell receptors are activated, cell killing begins.

T-helper cell A key cell of the immune system that decides whether a given foreign protein is "self" or "nonself" (harmless or harmful), and if the latter, activates a particular set of B-lymphocytes to manufacture antibodies against it.

toxins Proteins elaborated by some organisms that are poisonous to some other species.

tumor heterogeneity Differences in the cells composing a tumor or cancer.

Tumor Necrosis Factor A chemical, released by immune system cells, that can kill diseased cells, including cancers, but also produces side effects, such as loss of appetite and, in extreme instances, lowering of the blood pressure to shock levels and death.

Author's Note and Acknowledgments

When, in his eighties, world-famous physicist Hans Bethe confided to a friend that he was writing a personalized account of the creation of the first atomic bomb, his friend asked why, after so many years and so much controversy, he felt compelled to do it now. "I'm writing it for God," Bethe allegedly proclaimed. Gently the friend protested, "Hans, don't you think God already knows the story?" Whereupon Bethe shot back, "He doesn't know *my* version!"

This, then, is one version (of what doubtless will be many) of some of the major discoveries that have led to a more complete understanding of the nature of human cancer and better ways to prevent or treat it. In his advice to all writers, the great French novelist Stendhal urged, "Above all, be clear!" and, above all, I have striven to be. My main regret in bringing this book to a close is that its format did not give me the leeway to describe the exciting research programs of more Wistar Institute professors, such as Giorgio Trichieri with Natural Killer cells, Alonzo Ross with growth-factor receptors, Davor Solter with cell-nucleus transfers, and many others who so generously shared their knowledge with me. Unlike most independent biomedical research centers of comparable stature, The Wistar Institute lacks a substantial endowment fund and so must constantly seek other sources to meet its operating costs. That it does so successfully, year after year, without generating too much in the way of internal anxiety or staff turnover is due largely to the efforts of its associate director, Dr. Warren Cheston, a theoretical physicist by training and so well versed in creating order out of chaos; its director of development, Kurtis Meyer; its finance officer, Larry Keinath, and its board of managers, currently chaired by industrialist Robert A. Fox. I am especially grateful to public affairs director Diana Burgwyn for the innumerable ways she helped in the planning and researching of this book; her friendship already stands as one of its richest rewards. Knowing Diana to be as proficient a writer as she is a publicist, I would not be surprised if she were someday to produce her own account of the uniqueness of The Wistar Institute.

It is a rare commercially written book these days that runs a frictionless course from inception to completion in a form satisfactory to both author and publisher, and *Cell Wars* is no exception. Among the many individuals who helped cushion the jolts, I would like to pay special tribute to my peerless secretary, Wilma Kingsley; my friends Kenneth Kay, Vivian Gottlieb, and Roger Hirson; my literary agent, Mel Berger; my editors, Arthur Wang and Bill Newlin; and the man who enabled this work to survive its most perilous moments, my publisher, Roger Straus.

I humbly thank them all.

MARSHALL GOLDBERG, M.D.